# LEED® GREEN ASSOCIATE™

## V5 EXAM PREP

# PRACTICE TESTS & SUMMARY SHEETS

THIRD EDITION

(LEED GREEN ASSOCIATE EXAM PREPARATION GUIDE SERIES)

A. Togay Koralturk

Projectific

Published by:
Projectific, Inc.
8 The Green
Suite #4445
Dover, DE 19901
www.projectific.com

ISBN: 978-1-972631-07-2 (paperback)

Printed in the United States of America.

ATTENTION CORPORATIONS, UNIVERSITIES, COLLEGES, AND PROFESSIONAL ORGANIZATIONS. Special quantity discounts are available on bulk purchases of this book. Desk copies are also available for universities and academic institutions. For more information, please contact: customercare@projectific.com

Visit us on the Web: projectific.com

# CONTENTS

## PART 1 PRACTICE TESTS

# PART 2 SUMMARY SHEETS

## SECTION 5: SUMMARY SHEETS

# ALSO BY A. TOGAY KORALTURK

# SECTION 1:

# ABOUT THE LEED GREEN ASSOCIATE EXAM

## EXAM CONTENT

The LEED Green Associate exam contains 100 randomly delivered multiple-choice questions and must be completed in 2 hours. The exam questions reflect Knowledge Domains aligned with the LEED v5 rating system credit categories.

Below is the breakdown of the exam content areas and the number of questions per category:

- LEED Process (15 questions): USGBC/GBCI fundamentals, LEED rating system structure, certification benefits, available resources, rating system families, development process, category goals, impact areas, and certification process.
- Integrative Process Planning and Assessments (6 questions): Early assessment, systems thinking, life-cycle cost analysis, and roles of integrative project team members.
- Location and Transportation (9 questions): Site selection, transportation demand management, active and alternative transportation, VMT reduction, and equitable development.

- Sustainable Sites (8 questions): Site assessment, environmental considerations, open space, habitat conservation, heat island reduction, light pollution reduction, and rainwater management.
- Water Efficiency (10 questions): Outdoor and indoor water use reduction, graywater and rainwater use, smart irrigation, efficient fixtures and appliances, leak detection, and water metering.
- Energy and Atmosphere (15 questions): Building loads, energy efficiency, energy modeling, passive strategies, renewable energy, green power, energy monitoring, commissioning, and decarbonization.
- Materials and Resources (11 questions): Life-cycle impacts, embodied carbon, material selection, product declarations, VOCs, reuse, recycling, repurposing, and waste management.
- Indoor Environmental Quality (11 questions): Indoor air quality, low-emitting materials, ventilation, lighting, acoustics, thermal comfort, biophilic design, accessibility, and inclusion.

# LEED GREEN ASSOCIATE EXAM POLICIES AND PROCEDURES

In this section, we will provide details on the LEED Green Associate exam policies, procedures, and other important aspects you should know before taking your exam. However, as GBCI's policies and procedures are subject to change, it is also recommended that you visit USGBC's website and download the latest LEED Green Associate Candidate Handbook.

## HOW TO REGISTER FOR THE EXAM?

To register for the LEED Green Associate exam, you need to complete the following steps:

1) Create a USGBC site user account at www.usgbc.org/credentials, or log in if you already have one. **When filling out the required information, make sure that the name you enter exactly matches your name on the ID you will present to take the exam**. Any discrepancy may disqualify you from taking the exam.

2) Under the *Credentials* section, select the LEED Green Associate credential and follow the instructions to complete the application. Full-time and part-time students can take advantage of discounted rates for the LEED Green Associate exam. When submitting your exam registration, you will be required to enter your school, student ID number, and graduation date to receive the student discount.

3) Once your application is submitted and payment is processed, you will receive an Eligibility ID (EID) and will be redirected to prometric.com/gbci to schedule your exam date and location. Alternatively, you may choose to take the exam online through remote proctoring. Your EID is valid for 12 months from the date of registration.

4) When the exam is scheduled, you will receive a confirmation number onscreen and an email through Prometric. Keep your confirmation number, as you will need it to confirm, cancel, or reschedule your appointment.

5) Once you register and pay for your exam, you have one year to schedule and take your exam session. You may request one six-month extension of this one-year period.

## WHERE TO TAKE THE EXAM?

The LEED Green Associate exam can be taken at Prometric testing centers around the world, or you can take the exam online, in the comfort of your home or office, while being monitored by an offsite proctor. In the following sections, we will discuss the policies and procedures for both the testing center and the online proctored test.

## TESTING CENTER POLICIES AND PROCEDURES

Test-takers who choose to take their exam at a Prometric testing center should arrive 30 minutes before their scheduled appointment to sign in and present their identification, Eligibility ID (EID), and the confirmation number, which are emailed upon scheduling the appointment. Test-takers who arrive later than their scheduled exam time will lose their seat.

When signing in, test-takers must present a valid, non-expired, government-issued photo ID that also contains a signature. Once more, your name on the identification must exactly match the name on your exam registration. In case of a discrepancy, you may not be permitted to take your exam.

Nothing can be brought to the testing area. Small lockers may be available at the testing center to store your wallet, cell phone, keys, and other small items. However, laptops, briefcases, or large purses should not be taken to the testing center.

Throughout the exam session, you should remain seated. If you experience an issue or need to take a break, raise your hand to notify test center staff. Note that the exam will not pause if you take a break, and you must get the test proctor's permission to leave the room.

Expect to see other individuals entering and exiting the testing area throughout your exam, as Prometric testing centers offer a variety of exams. If the testing center is in a busy area, be prepared for potential noise distractions.

## POLICIES AND PROCEDURES FOR ONLINE PROCTORED TEST

To take the LEED Green Associate exam online, you will need a computer equipped with a webcam, a stable internet connection, and a private, quiet space where no one will enter or pass through your testing area.

Before scheduling your exam, be sure to review the online proctoring requirements on the Prometric website at prometric.com/gbci. On that website, you will have the opportunity to run a system test to ensure your computer and internet connection meets all the necessary criteria for a smooth testing experience. Taking this step before scheduling your exam is crucial to avoid any technical issues that could interrupt or invalidate your exam session. If your system cannot pass the test, you may instead choose to take the exam at a testing center.

On exam day, you will need to check in for the online proctored test before your scheduled appointment. During the check-in process, you will need to run a system check and provide photos of yourself, your identification, and your testing space. Your identification must be a valid, non-expired, government-issued photo ID that also contains a signature.

Once you submit your photos, they will be reviewed by the offsite proctor. If there is any issue, the proctor will contact you. To speed up the check-in process, make sure that your desk and the area around you are clear of all personal and preparatory items, including books, notes, and other materials. Otherwise, the proctor may ask you to show all sides of your testing environment via your webcam.

Throughout your online exam session, you will be under surveillance via your webcam and microphone by the offsite proctor. If someone enters your testing space or if there are loud distractions, you risk having your exam terminated. For this reason, choose a testing space free of interruptions and background noise.

Once everything meets the requirements, the proctor will start your exam. You should remain seated in front of your webcam throughout the exam session. If you encounter any technical issue during your exam, the proctor will attempt to resolve it in real time.

### SPECIAL TESTING ACCOMODATIONS

Accommodations can be requested if you have a documented disability that would prevent you from taking the exam under normal testing conditions. Prometric complies with the provisions of the Americans with Disabilities Act (ADA). There is no additional charge for special accommodations.

## DURING THE EXAM

### EXAM TUTORIAL AND EXIT SURVEY

Before the exam, there will be a 10-minute tutorial to demonstrate the exam software. At the end of the exam, there will be a 10-minute optional exit survey. The time allocated for the tutorial and the survey is not included in your 2-hour examination time.

## THE EXAM

The LEED Green Associate exam contains 100 randomly delivered multiple choice questions that must be completed in 2 hours.

As the exam starts, you will be presented with one question at a time. For every question, you have the option to answer the question, answer and mark it for later review, or skip it. You have the flexibility to move back and forth among the questions at any point and change your answer to any question you have previously addressed. The exam software will also list all the questions you have either skipped or marked for review so that you can easily revisit them.

**Make sure you read all the questions and the choices very carefully. If a question seems to have more than one answer, make sure you thoroughly understand the question and pay special attention to the wording. It doesn't matter how well you know the exam content; if you don't read carefully, the exam can easily trick you into selecting the wrong answer.**

Answer every single question since leaving a question blank will not award any points. If you are unsure about the answer to a question and are likely to spend extra time deciding on the correct answer, mark the question and come back to it later. Often, you will find it easier to answer such questions upon returning to them with a fresh perspective. If you forget something and struggle to answer a question, the questions or answer choices you see later in the exam can help you remember the right answer to a previous question.

At the end of the exam, make sure you use any extra time to review your answers, starting with the questions you felt most unsure about.

# AFTER THE EXAM

## EXAM RESULTS

After the exam, your exam score will be displayed on screen, and a score report will be emailed to you following your exam session. Within 72 hours of your appointment, your exam results will be processed and your Credentials Dashboard will be updated.

Note that candidates who take the LEED v5 beta exam (April–August 2026) will not receive immediate results; beta exam scores are expected to be released in October 2026.

## PASSING THE EXAM

As soon as you have passed the exam, you can use the title "LEED Green Associate" and/or the logo. Please remember that the abbreviation "LEED GA" is not approved by USGBC and should not be used.

If a LEED AP with specialty credential is earned after becoming a LEED Green Associate, your LEED Green Associate credential will be replaced by the LEED AP with specialty credential.

## FAILING THE EXAM

The full exam fee will be charged for each scheduled exam session after failing an exam. After three unsuccessful attempts, you must wait 90 days before submitting a new registration to GBCI.

## HOW TO MAINTAIN YOUR CERTIFICATION

All LEED Green Associates must earn 15 continuing education (CE) hours within 2 years of earning their credential, while all LEED APs with specialty must earn 30 continuing education (CE) hours within 2 years of earning their credential.

# SECTION 2:

## PRACTICE TEST 1

**It is recommended to study the Summary Sheets and Glossary before taking the practice tests.**

**1)** A project team selects low-emitting flooring materials that also have environmental product declarations (EPDs) and health product declarations (HPDs). Which credit categories does this single product selection support?

a) Water Efficiency and Indoor Environmental Quality
b) Energy and Atmosphere and Indoor Environmental Quality
c) Materials and Resources and Energy and Atmosphere
d) Indoor Environmental Quality and Materials and Resources

**2)** Which level of LEED BD+C certification can a project receive by meeting all the Minimum Program Requirements (MPRs), meeting all the prerequisites except one, and additionally earning 75 points?

a) Silver
b) Gold
c) Platinum
d) Cannot receive LEED certification

**3)** Which of the following strategies is not addressed in the Materials and Resources credit category?

a) Cradle-to-Cradle certified products

b) Providing storage for durable goods such as e-waste and furniture

c) Developing a construction waste policy

d) Purchasing renewable energy certificates (RECs)

**4)** Which of the following evaluates all the environmental effects of a product quantitatively for the whole lifetime of that material?

a) Life-cycle assessment (LCA)

b) Life-cycle costing (LCC)

c) Embodied energy

d) Triple bottom line

**5)** A mixed-use LEED v5 BD+C: New Construction project contains 65% school space and 35% warehouse space by gross floor area. Which project type requirements must the project meet?

a) The school project type requirements only

b) The warehouse and distribution center project type requirements only

c) Both the school and the warehouse and distribution center project type requirements

d) Neither set of project type requirements

**6)** Installing photovoltaic (PV) panels on the roof of a building, can contribute to which of the following? (Choose 2)

a) Generation of on-site renewable energy

b) Reducing heat island effect

c) Reducing stormwater runoff

d) Reducing building's heating loads

**7)** During the design phase, the commissioning provider (CxP) reviews the owner's project requirements (OPR) and the basis of design (BOD) and identifies a conflict between the two documents. What should the CxP do?

a) Revise the BOD to match the OPR without consulting the design team

b) Document the conflict and bring it to the attention of the owner and design team for resolution

c) Ignore the conflict because it will be resolved during construction

d) Defer to the BOD because it is the architect's responsibility

**8)** Which of the following represents a tradable, nontangible commodity associated with the qualities of renewable energy generation, and is a proof that when purchased, an amount of energy was created using renewable energy sources?

a) Renewable energy certificate (REC)
b) Carbon offset
c) Carbon footprint
d) Carbon neutrality

**9)** Which of the following is not a type of reflectivity measurement?

a) Solar reflectance index (SRI)
b) Solar reflectance (SR)
c) Albedo
d) Spatial daylight autonomy (sDA)

**10)** Which of the following would cause the largest increase in the outdoor water consumption of a project's landscaping?

a) Drip irrigation systems
b) Xeriscaping
c) Native plants
d) Turfgrass

**11)** Embodied energy of a product refers to the energy consumed during the:

a) Extraction of a product
b) Extraction and manufacturing of a product
c) Transportation of a product
d) Entire life-cycle of a product

**12)** Which of the following statements about Forest Stewardship Council (FSC) is false?

a) FSC is a voluntary program that sets the standards for wood product manufacturers
b) FSC aims to ensure responsible forest management in order to prevent deforestation and habitat
c) FSC provides chain of custody (CoC) certification
d) Projects seeking LEED certification should only use FSC certified wood products, or a local equivalent

**13)** When addressing the LEED requirements for the heat island effect, which of the following values do projects need to use for the non-roof measures? (Choose 2)

a) Solar reflectance (SR) value
b) Aged solar reflectance (SR) value
c) Solar reflectance index (SRI) value
d) Aged solar reflectance index (SRI) value

**14)** A facility manager wants to understand which building systems consume the most water in order to prioritize efficiency upgrades. Which tool provides this breakdown?

a) Energy use intensity (EUI) report
b) Water-end-use profile
c) Life-cycle assessment (LCA)
d) ENERGY STAR Portfolio Manager score

**15)** Minimizing the building footprint of a project, cannot aid in:

a) Reducing the heat island effect
b) Reducing the stormwater runoff
c) Increasing the amount of captured rainwater from the roof
d) Protecting the habitat

**16)** A site is landscaped so that it supports a diverse and balanced community of plants, animals, and microorganisms, using species that thrive without irrigation, pesticides, or fertilizers. What does LEED call such an environment?

a) Greenfield
b) Healthy habitat
c) Constructed ecosystem
d) Development footprint

**17)** A product that does not contain any recyclable content, is most similar to which of the following?

a) Closed system
b) Cradle-to-gate
c) Cradle-to-cradle
d) Cradle-to-grave

**18)** A building has a solar reflectance index (SRI) value of 82 on its roof and an albedo of 0.35 on its parking surfaces. Which statement best describes the relationship between these two measurements?

a) SRI and albedo measure the same property — the higher the value, the more heat is reflected
b) Albedo measures only solar reflectance, while SRI combines both solar reflectance and thermal emissivity to indicate a surface's ability to stay cool
c) SRI applies only to roofing materials, while albedo applies only to ground-level surfaces
d) A high SRI value always corresponds to a high albedo value

**19)** Which of the following abbreviations is disapproved by USGBC, and cannot be used under any circumstances?

a) LEED GA
b) USGBC
c) U.S. Green Building Council
d) Legacy LEED AP

**20)** Which rating system is eligible for a LEED precertification?

a) LEED BD+C: New Construction and Major Renovations
b) LEED BD+C: Core and Shell Development
c) LEED BD+C: Schools
d) LEED ID+C: Commercial Interiors

**21)** How many total points are available in the LEED ID+C rating systems?

a) 90 points
b) 95 points
c) 100 points
d) 110 points

**22)** Which of the following team member's primary role is to support and encourage the design integration required by LEED to streamline the application and certification process?

a) Facility Manager
b) LEED AP BD+C team member
c) LEED Green Associate
d) LEED Project Administrator

**23)** A building in a hot climate has a concrete structure with high thermal mass. How does this property benefit the building's energy performance?

a) It increases the building's solar reflectance, reducing cooling loads
b) It absorbs heat during the day and releases it slowly, reducing peak cooling demand and shifting loads to off-peak hours
c) It eliminates the need for mechanical cooling entirely
d) It reduces the building's embodied carbon by storing energy

**24)** Which of the following is used to label indoor plumbing fixtures?

a) Green-e
b) MERV
c) WaterSense
d) EPA

**25)** A building has both an energy-recovery ventilation system and demand-controlled ventilation. During winter, the building experiences high occupancy in a conference room. How do these two systems work together?

a) The demand-controlled ventilation shuts off the energy-recovery system to save energy
b) The demand-controlled ventilation increases outdoor air intake based on CO2 levels, while the energy-recovery system captures heat from the exhaust air to precondition the incoming cold fresh air, reducing heating energy
c) The energy-recovery system replaces the need for demand-controlled ventilation entirely
d) Both systems operate independently with no interaction

**26)** Faucets are rated for water efficiency using which unit of measurement?

a) Gallons per flush (gpf)
b) Gallons per minute (gpm)
c) Gallons per hour (gph)
d) Gallons per cycle (gpc)

**27)** By continuously reusing and recycling building materials instead of discarding them, a project team reduces the need to extract new raw materials. What LEED concept does this practice support?

a) Dematerialization
b) Source reduction
c) Material circularity
d) Life-cycle costing

**28)** Which of the following are administered by GBCI? (Choose 2)

a) LEED Reference Guides
b) LEED Professional Exams
c) LEED building certification process
d) LEED educational programs

**29)** A LEED O+M project was certified three years ago. The facility manager has not initiated recertification. What is the status of the project's LEED certification?

a) The certification remains valid indefinitely
b) The certification has expired and the project must pursue full certification again from scratch
c) The certification has expired because LEED O+M projects must recertify every three years
d) The certification is still valid because recertification is optional

**30)** Employing sustainable principles for the development of which of the following will make the biggest contribution to decreasing greenhouse gas emissions?

a) Built environments
b) Buildings
c) Roads
d) Single-occupancy vehicles

**31)** In an appropriate climate, using natural ventilation in a building can make a positive contribution to: (Choose 3)

a) Energy usage of the building
b) Indoor air quality of the building
c) Light pollution reduction
d) Heat island effect reduction
e) Reduction in the greenhouse gas emissions

**32)** A project team is debating whether to deconstruct or demolish an existing building on their site before new construction begins. Which approach does LEED favor?

a) Demolition
b) Deconstruction
c) Adaptive reuse
d) Demolition or deconstruction

**33)** Which of the following is the systematic process for analyzing and optimizing building system performance during the building operation phase?

a) Commissioning
b) Retrocommissioning
c) Life-cycle Assessment (LCA)
d) Systems thinking

**34)** Which of the following is the mission statement of USGBC?

a) To transform the way buildings and communities are designed, built and operated, enabling an environmentally and socially responsible, healthy and prosperous environment that improves the quality of life
b) Buildings and communities will regenerate and sustain the health and vitality of all life within a generation
c) A voluntary, consensus based rating system that provides third party verification that a building or community was designed and built to establish energy savings, water efficiency, location efficiency, improved indoor environmental quality, and stewardship of resources and sensitivity to their impacts
d) Recognize environmental leadership in the building industry

**35)** After the LEED certification preliminary review, how many appeals can a project file?

a) 5
b) 10
c) 20
d) Unlimited

**36)** A contractor orders 15% more drywall than the project specifications require to account for potential damage and cutting waste during installation. Most of the surplus is never used and is eventually discarded. What is this surplus called?

a) Commingled waste
b) Clean waste
c) Overage
d) Preconsumer recycled content

**37)** Which of the following construction-phase strategies reduces dependence on fossil fuels while also improving air quality for workers on-site?

a) Implementing an erosion and sedimentation control plan

b) Using battery-powered equipment instead of diesel-powered generators and machinery

c) Installing temporary MERV 13 filters in the construction trailer

d) Scheduling construction activities during cooler months to reduce energy demand

**38)** To address indoor water use, LEED requires all projects to meet minimum efficiency requirements for fixtures and fittings by complying with the maximum specified flush and flow rates or by reducing the project's aggregate water use by at least __________ through the use of more efficient fixtures and fittings.

a) 10%

b) 20%

c) 30%

d) 40%

**39)** A project team is deciding between two stormwater management strategies. Strategy A uses bioswales to slow and filter runoff. Strategy B uses dry ponds to temporarily detain stormwater. What is the key difference between these two approaches?

a) Bioswales hold water indefinitely while dry ponds drain within hours

b) Dry ponds use vegetation for filtration while bioswales rely on engineered basins only

c) Bioswales use a combination of engineered basins, soils, and vegetation to filter runoff, while dry ponds simply detain and slowly release water without filtration

d) Both strategies function identically and can be used interchangeably

**40)** Water that is consumed by the mechanical systems in buildings is called the:

a) Blowdown

b) Reclaimed water

c) Process water

d) Make-up water

**41)** Which of the following strategy is not a part of the Indoor Environmental Quality credit category?

a) Testing for radon or other contaminants inside the building

b) Conducting a flush-out

c) Participation in a demand response program

d) Providing daylighting inside the building

**42)** Which of the following strategies cannot address both daylighting, and quality views?

a) Windows configuration

b) Installing low cubicle partitions

c) Installing light tubes

d) Installing vision glazing

**43)** Which of the following strategies can decrease the stormwater runoff and reduce the heat island effect?

a) Installing a vegetated roof
b) Installing a cool roof
c) Rainwater harvesting
d) Providing undercover parking

**44)** Which of the following are intense workshops that are generally held in the beginning of the project and during the project milestones, with the participation of the stakeholders, the project team and experts?

a) Charrettes
b) Stakeholder meetings
c) Team meetings
d) Small task groups

**45)** A project team wants to reduce the building's lighting power density. Which combination of strategies would most directly achieve this?

a) Installing occupancy sensors
b) Replacing operable windows with fixed glazing and adding task lighting
c) Using high-efficacy LED luminaires and maximizing daylight through light shelves and skylights
d) Installing a building automation system and increasing the R-value of the roof insulation

**46)** When introducing natural daylighting into a building, each one of the following is true except:

a) Cooling loads of the building may increase
b) Productivity of the building occupants will increase
c) Electricity costs may be reduced with less usage of artificial lighting
d) Daylighting can help decrease glare

**47)** In LEED, building exterior is defined as:

a) Everything inside the waterproofing membrane
b) Everything outside the waterproofing membrane inclusive of the waterproofing membrane
c) Everything outside the waterproofing membrane exclusive of the waterproofing membrane
d) Everything outside the last layer of the facade

**48)** Which of the following cannot be considered as a source of biofuel?

a) Animal waste
b) Wood waste
c) Natural gas
d) Agricultural crops

**49)** A developer argues that a project is successful because it was completed under budget and generated strong financial returns. However, the project displaced a low-income community and increased local air pollution. Which framework would best reveal the shortcomings of evaluating this project on financial performance alone?

a) Life-cycle costing
b) Systems thinking
c) Triple bottom line (the three pillars of sustainability)
d) Life-cycle assessment

**50)** In order to establish an efficient lighting design, a project should address: (Choose 3)

a) Uplight
b) Glare
c) Backlight and light trespass on to adjacent sites
d) Reverberation time
e) Masking systems

**51)** Every project in any LEED rating system will be scored out of 100 points. However, there are also 10 additional bonus points. For the LEED BD+C rating systems, these points are:

a) 9 points from Project Priorities credit and 1 point from LEED AP credit
b) 6 points from Project Priorities credit and 4 point from LEED AP credit
c) 8 points from Project Priorities credit, 1 point from LEED AP credit, and 1 point from Innovation
d) 9 points from Project Priorities credit and 1 point from Regional Priority

**52)** A project team wants to reduce both sky glow and light trespass onto neighboring properties. Which single fixture specification would address both concerns simultaneously?

a) Fixtures with high lighting power density ratings
b) Fully shielded fixtures that direct all light downward
c) Fixtures with photosensors that dim based on daylight levels
d) Fixtures with high color temperature for maximum visibility

**53)** GBCI is the abbreviation for:

a) Green Business Certification Incentive
b) Green Building Certification Incentive
c) Green Building Certification Institute
d) Green Business Certification Inc.

**54)** A building uses operable windows for ventilation in mild weather and switches to mechanical cooling during extreme temperatures. What type of ventilation strategy is this?

a) Natural ventilation
b) Demand-controlled ventilation
c) Mixed-mode ventilation
d) Energy-recovery ventilation

**55)** In order to increase occupant comfort, a project should address: (Choose 3)

a) Thermal comfort
b) Building views
c) Ergonomics
d) Site assessment
e) Water efficiency

**56)** USGBC is testing a new compliance path that may be included in a future version of LEED. Project teams can optionally use this path in place of an existing credit requirement. What is this mechanism called?

a) Regional alternative compliance path
b) LEED interpretation
c) Pilot alternative compliance path
d) Pilot credit

**57)** LEED Reference Guides are published by:

a) GBCI
b) USGBC
c) GBCI and USGBC
d) LEED Council

**58)** Which of the following information cannot be found on the LEED scorecard (checklist)?

a) Point value of each credit
b) Names of the credits
c) Names of the credit's options (if applicable)
d) Yes / Maybe? / No

**59)** A project team using LEED v5 needs to upload prerequisite documentation, complete built-in credit calculators, and submit their application for GBCI review. Which platform should the team use?

a) LEED Online
b) ENERGY STAR Portfolio Manager
c) Arc
d) EDUCATION @USGBC

**60)** A project team completes a climate resilience assessment and identifies flooding and extreme heat as the two most relevant hazards to their site. The team then evaluates how a major flood event could damage the building's mechanical systems and displace occupants for months. What is the team evaluating in this second step?

a) The hazard assessment
b) The social determinants of health
c) The climate risk
d) The environmental justice implications

**61)** A LEED v5 project team sets a goal to reduce operational carbon by 40% below baseline over the building's first five years. Where should this goal first be recorded?

a) In the commissioning provider's report
b) In the LEED scorecard on the Arc platform
c) In the owner's project requirements (OPR)
d) In the construction waste management plan

**62)** Which of the following is the presentation of a product or a material as being more environmentally friendly than it actually is?

a) Greenwashing
b) Green chemistry
c) Green misleading
d) Green marketing

**63)** Which of the following is the most effective strategy to reduce solid waste?

a) Reusing materials
b) Source reduction
c) Waste-to-energy conversion
d) Recycling

**64)** What are the three components of the triple bottom line (the three pillars of sustainability)? (Choose 3)

a) Environmental stewardship
b) Social equity and justice
c) Systems thinking
d) Economic prosperity
e) Sustainable practices

**65)** Which of the following will aid in reducing outdoor water use? (Choose 3)

a) Native plants
b) Xeriscaping
c) Invasive plants
d) Turfgrass
e) Adapted plants

**66)** Exemplary performance points are awarded under the:

a) Project Priorities credit
b) Regional Priority credit
c) Pilot credits
d) Credits that qualify for exemplary performance

**67)** Minimum efficiency reporting value (MERV) rates the:

a) Efficiency of the HVAC equipment
b) Efficiency of the filtration media
c) Energy efficiency of an equipment
d) Water efficiency of indoor fixtures

**68)** A new hospital is being built in a low-income neighborhood that has historically lacked access to green spaces, clean air, and quality healthcare. The project team wants to address the root causes of health disparities in this community. Which concept should guide their approach?

a) Resilience
b) Social determinants of health
c) Passive survivability
d) Demand response

**69)** Which of the following statements is false?

a) An interest savings account can be an example of a positive feedback loop; as the account grows, more interest is earned which in turn brings further account growth
b) Open systems constantly consume other items, use them, and produce waste at the end
c) Closed systems do not produce any waste product at the end
d) Any action taken in a leverage point of the system will have no effect

**70)** Which of the following materials is an inherently nonemitting material?

a) Flooring adhesive
b) Untreated marble
c) Epoxy paint
d) Window sealant

**71)** Which of the following cannot be used to obtain graywater?

a) Water from toilets
b) Water from bathroom sinks
c) Water from bathtubs
d) Water from showers

**72)** Which of the following rating systems require in-field verification (site visit) in order to verify all the LEED requirements are met?

a) LEED O+M: Commercial Interiors
b) LEED Residential
c) LEED BD+C: New Construction and Major Renovations
d) LEED BD+C: Core and Shell Development
e) LEED BD+C: Healthcare

**73)** Which of the following LEED certification fees are flat fees, and are the same for both USGBC members and non-members? (Choose 2)

a) Review fees
b) Project registration fees
c) Appeal fees
d) Credit interpretation ruling (CIR)

**74)** A project team wants to identify whether any endangered species are present near their proposed site. Which two resources could they reference?

a) ENERGY STAR Portfolio Manager and Walk Score
b) U.S. Endangered Species Act and the IUCN Red List
c) American Bird Conservancy and ASHRAE 62.1
d) NatureServe and the EPA WaterSense Water Budget Tool

**75)** A project team designs a building using only the space and resources necessary to meet its functional requirements, avoiding excess in both size and material consumption. Which concept best describes this overall philosophy?

a) Material circularity
b) Dematerialization
c) Sufficiency
d) Adaptive reuse

**76)** Integrative process starts from the:

a) Pre-design phase
b) Design phase
c) Construction phase
d) Building operation phase

**77)** Which of the following is the person that will be responsible for running the building operations after the completion of construction?

a) Stakeholder
b) Facility manager
c) Mechanical Engineer
d) Project administrator

**78)** Which of the following individuals or entities can qualify as the commissioning provider (CxP) under LEED's fundamental commissioning requirements? (Choose 3)

a) Third-party consultant with no involvement in the project's design or construction
b) Owner's employee
c) Contractor's employee
d) Subcontractor's employee
e) Completely independent employee of the design or construction firm

**79)** In every case, the commissioning provider (CxP) in the project must directly report to the:

a) Design team
b) Construction team
c) Contractor
d) Owner

**80)** Which of the following statements about regenerative design is false?

a) Regenerative design is one step ahead of sustainable design
b) Regenerative design projects do not use any more energy than they can produce
c) Regenerative design does not contribute to the restoration of the lost environment and habitat
d) Regenerative design creates no waste

**81)** A project team commits to reaching net-zero emissions for its building's energy performance. According to the recommended sequence of decarbonization actions, what should the team address first?

a) Procuring off-site renewable energy
b) Reducing energy use through better design or energy retrofits
c) Purchasing carbon offsets
d) Installing energy storage

**82)** Which of the following refers to the understanding of each and every system of a building, while also understanding their relationships and looking at the project as a whole?

a) Feedback loop
b) Systems thinking
c) Leverage point
d) Linear approach

**83)** LEED requires all projects to create and implement an erosion and sedimentation control plan during construction. Which standard does LEED reference for this requirement in the United States?

a) ASHRAE Standard 62.1-2022
b) Institute of Transportation Engineers Parking Generation Manual
c) U.S. EPA 2022 Construction General Permit
d) ANSI Consensus National Standard Guide 2.0

**84)** In order to be eligible for a LEED BD+C or a LEED O+M rating system certification, a project should contain a minimum of:

a) 1,000 square feet (93 square meters) gross floor area
b) 2,000 square feet (186 square meters) gross floor area
c) 3,000 square feet (279 square meters) gross floor area
d) 4,000 square feet (372 square meters) gross floor area

**85)** By becoming a USGBC member, projects can pay reduced: (Choose 2)

a) Project registration fees
b) Project review fees
c) LEED interpretation fees
d) LEED Prototype fees

**86)** Urban development intensifies, attracting more businesses, which attract more residents, which drives further development. Which systems concept does this escalating cycle illustrate?

a) Negative feedback loop
b) Leverage point
c) Emergent property
d) Positive feedback loop

**87)** A project team incorporates live plants in the lobby, uses natural wood finishes throughout the corridors, and designs floor plans inspired by organic patterns found in nature. Which design approach does this represent?

a) Regenerative design
b) Inclusive design
c) Biophilic design
d) Adaptive reuse

**88)** LEED requires all projects to protect which type of vegetation on the construction site?

a) All existing vegetation regardless of status
b) Only vegetation listed in the IUCN Red List
c) 100% of special-status vegetation as defined by local, state, or federal entities
d) Only native plants that were present before construction began

**89)** Which of the following defines prime farmlands?

a) U.S. Department of Agriculture U.S. Code of Federal Regulations Title 7, Volume 6
b) U.S. Environmental Protection Agency (EPA), National Priority List
c) U.S. Housing and Urban Development
d) U.S. Department of Treasury, Community Development Financial Institutions Fund

**90)** Which of the following is not an alternative water source?

a) Collected rainwater
b) Captured condensate
c) Rejected water from reverse osmosis systems
d) Water from the public utilities

**91)** Which of the following causes glare?

a) Uplight
b) High-angle front lights
c) Light trespass
d) Backlight

**92)** Which of the following provide in-field verification to every LEED Residential project?

a) Home Raters
b) Green Raters
c) Green Classroom Professional (GCP)
d) LEED Green Associates

**93)** Thermal comfort is defined by a combination of six factors. Which of the following is not one of them?

a) Air movement
b) Air temperature
c) Metabolic rate
d) Space volume

**94)** Which of the following is not a phase of the LEED certification process?

a) Project registration
b) Review
c) Certification or denial
d) Appeal

**95)** __________ hold the excess rainwater for some time, thereby allowing the rainwater to slowly seep into the ground without contamination. __________ is/are excavated area(s) that detain and slow down stormwater, but are dry at other times.

a) Dry ponds
b) Bioswales
c) Low-impact developments
d) Green infrastructure

**96)** Which of the following is the least effective irrigation system to address outdoor water use reduction?

a) Bubbler distribution systems
b) Drip irrigation systems
c) Spray irrigation
d) Channel water directly to root systems

**97)** Which of the following cannot be a strategy to reduce greenhouse gases resulting from the burning fossil fuels?

a) Allowing telecommuting of employees
b) Allowing for compressed workweek calendar
c) Low-impact development (LID)
d) Using electric vehicles

**98)** Which of the following indicates a material's ability to stay cool by reflecting solar radiation and emitting thermal radiation, combining both reflectance and emissivity to rank the material?

a) Solar reflectance index (SRI)
b) Solar reflectance (SR)
c) Albedo
d) Heat island effect

**99)** Dark colored, non-reflective surfaces absorb heat during hot weather, and release it into the atmosphere; this releasing of heat is called:

a) Albedo
b) Emissivity
c) Heat island effect
d) Infrared emittance

**100)** Which of the following is not a goal of the Location and Transportation credit category?

a) Encourage development in dense areas
b) Encourage the use of electric vehicles
c) Encourage development near diverse uses
d) Encourage light pollution reduction

# PRACTICE TEST 1 – ANSWERS

**A score above 80 would indicate very well preparation for the exam.**

| | | | | | | | |
|---|---|---|---|---|---|---|---|
| 1 | D | 26 | B | 51 | A | 76 | A |
| 2 | D | 27 | C | 52 | B | 77 | B |
| 3 | D | 28 | B and C | 53 | D | 78 | A, B and E |
| 4 | A | 29 | C | 54 | C | 79 | D |
| 5 | A | 30 | A | 55 | A, B and C | 80 | C |
| 6 | A and B | 31 | A, B and E | 56 | C | 81 | B |
| 7 | B | 32 | B | 57 | B | 82 | B |
| 8 | A | 33 | B | 58 | C | 83 | C |
| 9 | D | 34 | A | 59 | C | 84 | A |
| 10 | D | 35 | D | 60 | C | 85 | A and B |
| 11 | D | 36 | C | 61 | C | 86 | D |
| 12 | D | 37 | B | 62 | A | 87 | C |
| 13 | A and B | 38 | B | 63 | B | 88 | C |
| 14 | B | 39 | C | 64 | A, B and D | 89 | A |
| 15 | C | 40 | C | 65 | A, B and E | 90 | D |
| 16 | B | 41 | C | 66 | A | 91 | B |
| 17 | D | 42 | C | 67 | B | 92 | B |
| 18 | B | 43 | A | 68 | B | 93 | D |
| 19 | A | 44 | A | 69 | D | 94 | D |
| 20 | B | 45 | C | 70 | B | 95 | A |
| 21 | D | 46 | D | 71 | A | 96 | C |
| 22 | B | 47 | B | 72 | B | 97 | C |
| 23 | B | 48 | C | 73 | C and D | 98 | A |
| 24 | C | 49 | C | 74 | B | 99 | C |
| 25 | B | 50 | A, B and C | 75 | C | 100 | D |

# PRACTICE TEST 1 ANSWERS & EXPLANATIONS

**1) Correct answer:** D. Indoor Environmental Quality and Materials and Resources.

Low-emitting flooring directly supports Indoor Environmental Quality by reducing occupant exposure to volatile organic compounds (VOCs). The same product's EPD and HPD contribute to Materials and Resources credit category, which rewards products with transparent environmental and health disclosures.

**2) Correct answer:** D. Cannot receive LEED certification.

In order to achieve a LEED certification: every project must comply with the Minimum Program Requirements (MPRs) regardless of the rating system. Then there are prerequisites of each rating system that are mandatory to achieve for LEED certification. Then, the project team must choose credits to pursue and collect more than 40 points. All the MPRs and prerequisites are mandatory, however projects are free to choose the credits they want to pursue.

**3) Correct answer:** D. Purchasing renewable energy certificates (RECs).

Purchasing renewable energy certificates (RECs) is addressed under the **Energy and Atmosphere** credit category.

**4) Correct answer:** A. Life-cycle assessment (LCA).

**Life-cycle assessment (LCA)** evaluates all the environmental effects of a product quantitatively for the whole lifetime of that material. **Life-cycle costing** assesses a products total cost for the whole lifetime of the product by evaluating both the initial price and the operating costs. **Embodied energy** is the total energy consumed resulting from a product's manufacturing, transportation, installation, and use. **Triple bottom line** represents the 3 components that every green building should establish: economic prosperity (profit), social responsibility (people) and environmental stewardship (planet).

**5) Correct answer:** A. The school project type requirements only.

Some prerequisites and credits define specific requirements for specific project types, such as healthcare, schools, warehouses and distribution centers, or residential projects. For example, the No Smoking or Vehicle Idling prerequisite prohibits smoking on the entire site for school projects, while other projects may provide designated outdoor smoking areas. For mixed-use projects, project type requirements must be met if that project type comprises 60% or more of the project's gross floor area. In this question, the school space comprises 65% of the gross floor area, so the school project type requirements must be met.

**6) Correct answers:** A. Generation of on-site renewable energy. B. Reducing heat island effect.

Photovoltaic (PV) panels generate on-site renewable energy, and are categorized as active solar power systems. Because the PV panel will absorb the solar heat and convert it into electricity, this will also result in the reduction of the heat island effect. However, PV panels cannot make any contribution to reducing stormwater runoff, or the building's heating loads.

**7) Correct answer:** B. Document the conflict and bring it to the attention of the owner and design team for resolution.

The commissioning provider acts as the owner's advocate and is responsible for identifying conflicts between the OPR and BOD. The OPR documents the owner's vision and performance requirements, while the BOD describes how the design team intends to meet those requirements. When a conflict exists, the CxP must document it and facilitate resolution between the owner and design team — not unilaterally make changes or ignore the issue. This is a critical part of the commissioning process that ensures the building is designed and built as truly intended.

**8) Correct answer:** A. Renewable energy certificate (REC).

**Renewable energy certificates (RECs)** are tradable, intangible instruments representing the environmental and other non-power attributes of renewable electricity generation. One REC generally represents 1 megawatt-hour (MWh) of electricity generated from an eligible renewable resource and delivered to the grid.

**Carbon offset** is a reduction of carbon dioxide (CO2) made in order to compensate, or offset an equivalent carbon dioxide (CO2) emission made elsewhere.

**Carbon footprint** refers to the measure of greenhouse gas emissions associated with an activity, such as construction, operation, transportation, or construction materials.

**Carbon neutrality**, is to emit no more carbon emissions than can realistically be offset.

**9) Correct answer:** D. Spatial daylight autonomy (sDA).

**Spatial daylight autonomy (sDA)** is a metric used to describe the annual sufficiency of ambient daylight in building interiors and is not a reflectivity measurement. **Solar reflectance index (SRI)**, **solar reflectance (SR)** and **albedo** are all reflectivity measurements.

**10) Correct answer:** D. Turfgrass.

**Turfgrass** requires regular irrigation and needs a high amount of irrigation water.

**Drip irrigation** system is the most water efficient irrigation system, and therefore would reduce the outdoor water consumption. **Xeriscaping** is a type of landscaping that does not need any irrigation. **Native plants** (indigenous plants) are the type of plants that occur and develop naturally in a specific location, and can thrive without much maintenance. Therefore using native plants will also reduce the outdoor water consumption.

**11) Correct answer:** D. Entire life-cycle of a product.

Embodied energy represents the total energy consumed resulting from a product's manufacturing, transportation, installation, use, and disposal.

**12) Correct answer:** D. Projects seeking LEED certification should only use FSC certified wood products, or a local equivalent.

Projects seeking LEED certification can use wood products which are not FSC certified.

**13) Correct answers:** A. Solar reflectance (SR) value. B. Aged solar reflectance (SR) value.

In LEED credit calculations (under the Heat Island Reduction credit), solar reflectance (SR) and aged SR values are used for the non-roof measures, and solar reflectance index (SRI) and aged SRI values should be used for roofs.

**14) Correct answer:** B. Water-end-use profile.

A **water-end-use profile** identifies the most significant uses of water in a project by breaking down consumption by system or fixture type — such as cooling towers, irrigation, restrooms, and kitchen facilities. This enables facility managers to target the largest water consumers for efficiency improvements. An **anergy use intensity (EUI) report** addresses energy, not water. A **life-cycle assessment (LCA)** evaluates environmental impacts of materials. **ENERGY STAR Portfolio Manager** benchmarks overall energy performance but does not provide a detailed water end-use breakdown.

**15) Correct answer:** C. Increasing the amount of captured rainwater from the roof.

Minimizing building footprint will reduce the amount of captured rainwater from the roof, since the area of the roof will also be reduced.

**16) Correct answer:** B. Healthy habitat.

A healthy habitat is an environment that supports a diverse and balanced community of plants, animals, and microorganisms. Healthy habitats often rely on native and adapted plant species that survive without irrigation, pesticides, or fertilizers, and as the climate changes, project teams should also consider future conditions when selecting plants. A greenfield is simply land that has not been previously developed — it says nothing about the health of its ecosystem.

**17) Correct answer:** D. Cradle-to-grave.

A product that does not contain any recyclable content will eventually end up in the landfills. This is most similar with **cradle-to-grave** approach, as it investigates the materials from their extraction to their disposal, because cradle-to-grave products only have one life-cycle.

Unlike cradle-to-grave, **closed systems** are sustainable systems that do not produce any waste product at the end, thereby they're similar with cradle-to-cradle. **Cradle-to-cradle** evaluates materials to have infinite life-cycles through recycling to form a closed system.

**Cradle-to-gate** assessment evaluates a product's partial life-cycle from its resource extraction/harvesting to becoming a manufactured product ready for sales at the factory gate, which is not relevant.

**18) Correct answer:** B. Albedo measures only solar reflectance, while SRI combines both solar reflectance and thermal emissivity to indicate a surface's ability to stay cool.

Albedo measures how much solar energy a surface reflects on a scale of 0 to 1. SRI goes further by combining both solar reflectance and thermal emissivity to provide a more comprehensive measure of a material's ability to reject solar heat and stay cool. A material can have moderate reflectance but high emissivity, resulting in a high SRI. The two metrics are related but not interchangeable, and neither is restricted to specific surface types.

**19) Correct answer:** A. LEED GA.

LEED GA abbreviation is disapproved by USGBC and should not be used.

**20) Correct answer:** B. LEED BD+C: Core and Shell Development.

LEED BD+C: Core and Shell Development projects are eligible for precertification.

**21) Correct answer:** D. 110 points.

In all the LEED rating systems there are 110 available points, however all rating systems are scored out of 100 points.

**22) Correct answer:** B. LEED AP BD+C team member.

The primary role of the LEED AP BD+C team member is to support and encourage the design integration required by LEED in order to streamline the application and certification process. Facility manager is the person that will be responsible for running the operations of the buildings after the completion of construction. LEED Green Associate team member will have the general knowledge of green building practices, and will provide support to other project team members. The LEED Project Administrator is a team member that will act as a project manager, overseeing the LEED certification process.

**23) Correct answer:** B. It absorbs heat during the day and releases it slowly, reducing peak cooling demand and shifting loads to off-peak hours.

Thermal mass refers to a material's ability to absorb and store thermal energy. High-density materials like concrete absorb heat during the day, reducing peak cooling loads, and release it slowly during cooler periods. This effectively shifts some of the cooling demand to off-peak hours, which can reduce energy costs and strain on the grid. Thermal mass does not affect solar reflectance (which is a surface property), cannot eliminate mechanical cooling on its own, and has no relationship to reducing embodied carbon.

**24) Correct answer:** C. WaterSense.

**WaterSense** is a program developed by the U.S. Environmental Protection Agency to identify high-performance, water-efficient fixtures and fittings.

**Green-e** is the leading certification program for the green power generation in the United States, and provides certification for green power, renewable energy certificates (RECs) and carbon offsets.

**Minimum efficiency reporting value (MERV)** rates the air filters according to their success in removing particles from the air.

**EPA**, United States Environmental Protection Agency, developed the WaterSense label to identify high-performing fixtures.

**25) Correct answer:** B. The demand-controlled ventilation increases outdoor air intake based on CO2 levels, while the energy-recovery system captures heat from the exhaust air to precondition the incoming cold fresh air, reducing heating energy.

An energy-recovery ventilation system and a demand-controlled ventilation system create a powerful synergy. When the conference room occupancy is high, CO2 levels rise, triggering the demand-controlled ventilation to increase the outdoor air supply. Bringing in more cold outdoor air in winter would normally increase heating costs significantly. However, the energy-recovery ventilation system captures heat from the outgoing exhaust air and transfers it to the incoming fresh air, reducing the energy penalty of increased ventilation. This combination optimizes both indoor air quality and energy performance.

**26) Correct answer:** B. Gallons per minute (gpm).

Faucets are flow fixtures, and their water efficiency is measured in gallons per minute (gpm), which indicates how much water flows while the fixture is in use. Gallons per flush (gpf) applies to flush fixtures such as toilets and urinals. Gallons per hour and gallons per cycle are not standard LEED water efficiency metrics.

**27) Correct answer:** C. Material circularity.

**Material circularity** is a system in which resources are continuously reused and recycled, reducing dependence on virgin materials and minimizing waste. **Dematerialization** refers to reducing the total quantity of materials used. **Source reduction** involves sizing materials precisely to prevent waste. **Life-cycle costing** estimates total costs over a product's lifetime. The key distinction is that material circularity specifically addresses keeping materials in use rather than extracting new ones.

**28) Correct answers:** B. LEED Professional Exams. C. LEED building certification process.

LEED Reference Guides and educational programs are administered by USGBC.

**29) Correct answer:** C. The certification has expired because LEED O+M projects must recertify every three years.

LEED O+M certifications require recertification every three years to maintain their status.

**30) Correct answer:** A. Built environments.

The built environment refers to all the man-made surroundings that are needed for human activity, from roads, to buildings, to neighborhoods. The built environment contributes to 67% of all the greenhouse gas emissions. Therefore, employing sustainable principles for the development of built environments will make the biggest contribution to decreasing greenhouse gas emissions.

**31) Correct answers:** A. Energy usage of the building. B. Indoor air quality of the building. E. Reduction in the greenhouse gas emissions.

Using natural ventilation will reduce the energy needed for mechanical ventilation, and less energy usage will result in the reduction of the greenhouse gas emissions. Circulation of outside fresh air will result in increased indoor air quality of the building. However, heat island effect and light pollution are not relevant to natural ventilation.

**32) Correct answer:** B. Deconstruction.

LEED favors deconstruction over demolition because it enables significantly more building materials to be carefully disassembled and salvaged for reuse, rather than being destroyed and sent to landfill as mixed debris. Deconstruction supports material circularity and waste diversion goals.

**33) Correct answer:** B. Retrocommissioning.

**Commissioning (Cx)** is a systematic investigation by skilled personnel that compares building performance with the project goals, design specifications, and most importantly, the owner's project requirements (OPR). Commissioning takes place during the design and construction phases, however, retrocommissioning is conducted during the building operation phase. **Life-cycle assessment (LCA)** evaluates all the environmental effects of a product quantitatively for the whole lifetime of that material. **Systems thinking** refers to the understanding of each and every system of a building, while also understanding their relationships and looking at the project as a whole.

**34) Correct answer:** A. To transform the way buildings and communities are designed, built and operated, enabling an environmentally and socially responsible, healthy and prosperous environment that improves the quality of life.

USGBC's Mission: "To transform the way buildings and communities are designed, built, and operated, enabling an environmentally and socially responsible, healthy, and prosperous environment that improves the quality of life."

USGBC's Vision: "Buildings and communities will regenerate and sustain the health and vitality of all life within a generation."

**35) Correct answer:** D. Unlimited.

Projects can file unlimited number of appeals during the LEED certification process.

**36) Correct answer:** C. Overage.

**Overage** is extra material ordered as a precaution that often goes unused. LEED encourages minimizing overage through careful design and procurement. When surplus remains, project teams should return it to the manufacturer or donate it to a reuse outlet rather than sending it to a landfill.

**Commingled waste** is single-streamed recycling.

**Clean waste** is nonhazardous construction debris.

**Preconsumer recycled content** is manufacturing by-products redirected into new products before reaching a consumer.

**37) Correct answer:** B. Using battery-powered equipment instead of diesel-powered generators and machinery.

Battery-powered equipment eliminates the burning of diesel fuel on-site, which simultaneously reduces fossil fuel dependence and improves air quality for construction workers by removing exhaust fumes, particulate matter, and combustion by-products from the work environment.

**38) Correct answer:** B. 20%.

To address indoor water use, LEED requires all projects to meet minimum efficiency requirements for fixtures and fittings by complying with the maximum specified flush and flow rates or by reducing the project's aggregate water use by at least 20% through the use of more efficient fixtures and fittings.

**39) Correct answer:** C. Bioswales use a combination of engineered basins, soils, and vegetation to filter runoff, while dry ponds simply detain and slowly release water without filtration.

Bioswales are stormwater control features that use engineered basins, soils, and vegetation to both slow down and filter runoff as it passes through. Dry ponds (detention ponds) are excavated basins that temporarily hold and slow stormwater during rain events but remain dry at other times — they detain water but do not actively filter it through vegetation and soil media the way bioswales do. Understanding the functional difference between these two common LID strategies is important for the exam.

**40) Correct answer:** C. Process water.

Water that is consumed by the mechanical systems in buildings is called the **process water.** To minimize the deposit of scales, some of the cooling tower's water is replaced by fresh potable water, and this removal of water is called **blowdown. Reclaimed water** is the former blackwater that has been treated and purified for reuse. **Make-up water** is used to replace both the lost water from evaporation and the blowdown.

**41) Correct answer:** C. Participation in a demand response program.

Participation in a demand response program is part of the **Energy and Atmosphere** credit category.

**42) Correct answer:** C. Installing light tubes.

Light tubes, also called sun tubes or sun pipes, are structures that are used to transport sunlight inside a building. To illustrate, imagine a big pipe with a reflective inside surface that goes through the roof, to the ceiling of a room. It captures the sunlight on the roof and transports it inside the pipe all the way down to the room. Therefore, light tubes will only provide daylighting, and will not provide quality views.

**43) Correct answer:** A. Installing a vegetated roof.

**Vegetated roofs** absorb and retain rainwater through their soil and plant material, reducing both the volume and peak flow of stormwater runoff. At the same time, they lower the heat island effect by reflecting more solar radiation than conventional dark roofs and cooling the surrounding air through evapotranspiration.

**Cool roofs** reflect significantly more sunlight and absorb less heat than traditional dark-colored roofs, however they cannot decrease the stormwater runoff since they will not absorb any rainwater.

**Rainwater harvesting** is a rainwater management strategy, which collects and filters the rainwater to be re-used as an alternative to potable water. The filtered rainwater can be used for irrigation, toilet and urinal flushing, or it can even be used as the process water of the cooling towers. Rainwater harvesting will decrease the stormwater runoff, however will not reduce the heat island effect.

**Providing undercover parking** will reduce the heat island effect, but will not affect the stormwater runoff. If a project provides underground parking instead of undercover parking, and increase the landscaping area, then the project would be able to address both the heat island effect and the stormwater runoff.

**44) Correct answer:** A. Charrettes.

**Charrettes** are intense workshops that are generally held in the beginning of the project and during the project milestones, with the participation of the stakeholders, the project team and experts.

**Stakeholder meetings** are conducted among the project team, stakeholders, neighbors, and community members in order to understand and discuss community needs, issues, and concerns.

**Team meetings** are helpful to discuss strategies among team members from different disciplines. For example, the design of a building envelope will have a great impact on the building's HVAC system. The level of insulation that a facade provides will affect the heating and cooling loads of the building, which will affect the size and type of the HVAC system to be used in the project. Thus, in a team meeting, the facade designer and MEP designer can work collaboratively to discover the best design solutions for both the facade and the HVAC system.

**Small task groups** are held inside each discipline and together with the involvement of consultants, if needed, to explore particular topics, research, and the refinement of ideas. Then, the results can be collaborated with the whole project team.

**45) Correct answer:** C. Using high-efficacy LED luminaires and maximizing daylight through light shelves and skylights.

Lighting power density measures the amount of power used per square foot when all lighting is turned on. High-efficacy LEDs consume less power per lumen than conventional fixtures, directly reducing installed lighting power density. Maximizing daylight through light shelves and skylights allows photosensors to dim or turn off electric lights, further reducing the effective lighting load. Fixed glazing does not improve daylighting over operable windows. Building automation and roof insulation address other building systems, not lighting power density specifically.

**46) Correct answer:** D. Daylighting can help decrease glare.

When introducing natural daylighting, cooling loads of the building may increase if technical skylights are not installed, which allow only the sunlight to enter a building, rather than the additional heat.

According to researches, natural daylighting is known to increase productivity, and health. With the use of natural daylighting, electricity costs will also be reduced with the less usage of artificial lighting.

However, a major problem that needs to be addressed when introducing daylighting is the glare it creates. Projects need to think about installing glare control devices when using daylighting.

**47) Correct answer:** B. Everything outside the waterproofing membrane inclusive of the waterproofing membrane.

In LEED, **building exterior** is defined as everything outside the waterproofing membrane inclusive of the waterproofing membrane. On the other hand, **building interior** is defined as everything inside the waterproofing membrane, exclusive of the waterproofing membrane.

**48) Correct answer:** C. Natural gas.

Biofuel-based resources include wood waste, landfill gas, agricultural crops or residues, animal waste, and other organic materials.

**49) Correct answer:** C. Triple bottom line (the three pillars of sustainability).

The **triple bottom line** (the **three pillars of sustainability**) evaluates project success across three dimensions: economic prosperity (profit), social equity and justice (people), and environmental stewardship (planet). While this project succeeded financially, it failed on the social dimension (displacing a community) and the environmental dimension (increasing pollution). Evaluating only profit ignores the other two pillars of sustainability. **Life-cycle costing** addresses financial costs over time. **Systems thinking** examines interrelationships. **Life-cycle assessment** (**LCA**) evaluates environmental impacts of products.

**50) Correct answers:** A. Uplight. B. Glare. C. Backlight and light trespass on to adjacent sites.

In order to establish an efficient lighting design, a project should address **backlight**, **uplight**, **glare** and **light trespass**. **Reverberation time** is the time span between a sound is produced and it dies away. **Masking systems**, or sound masking systems, are equipments used to reduce the background noise in spaces.

**51) Correct answer:** A. 9 points from Project Priorities credit and 1 point from LEED AP credit.

LEED projects are scored out of 100 base points, with up to 10 bonus points available: 9 under the Project Priorities credit and 1 under the LEED AP credit.

**52) Correct answer:** B. Fully shielded fixtures that direct all light downward.

Fully shielded fixtures eliminate uplight — light emitted above the horizontal plane that contributes to sky glow — by directing all light downward. This same downward orientation also helps prevent light trespass, which is unwanted light spilling onto adjacent properties. Photosensors reduce energy use by dimming lights when daylight is sufficient but do not control the direction of light. High lighting power density increases energy consumption. Color temperature affects light appearance, not direction. Fully shielded fixtures are the single specification that addresses both uplight and light trespass through directional control.

**53) Correct answer:** D. Green Business Certification Inc.

Green Business Certification Inc. (GBCI), formerly known as Green Building Certification Institute (GBCI), which changed the name on 16th of April 2015, is the sister organization of USGBC that administers the LEED building certification and LEED professional exams.

**54) Correct answer:** C. Mixed-mode ventilation.

Mixed-mode ventilation combines natural ventilation (operable windows) with mechanical systems, switching between them based on conditions. Pure natural ventilation relies solely on architectural features without mechanical backup. Demand-controlled ventilation adjusts mechanical outdoor air based on occupancy sensors. Energy-recovery ventilation uses exhaust air to precondition incoming fresh air but is entirely mechanical.

**55) Correct answers:** A. Thermal comfort. B. Building views. C. Ergonomics.

To increase occupant comfort, a project should address thermal comfort, building views, ergonomics, acoustics, and daylighting.

**56) Correct answer:** C. Pilot alternative compliance path.

Pilot alternative compliance paths (ACPs) are compliance paths that USGBC is considering for inclusion in future versions of LEED. They can optionally be used in place of existing LEED credits while USGBC evaluates their effectiveness. Regional ACPs address location-specific standards for international projects. LEED interpretations clarify existing credit requirements.

**57) Correct answer:** B. USGBC.

LEED Reference Guides are published by USGBC.

**58) Correct answer:** C. Names of the credit's options (if applicable).

The credit scorecard shows the names of the prerequisites and credits, the point value of each credit, and yes/maybe/no columns. Names of the credit's options cannot be found on the LEED scorecard. You can visit the USGBC's website to download a scorecard.

**59) Correct answer:** C. Arc.

**Arc** is the online platform for LEED v5 where project teams register projects, complete prerequisite and credit documentation using standardized forms with built-in calculators, and submit applications to GBCI for certification review. **LEED Online** was used for LEED v4 and v4.1. **ENERGY**

**STAR Portfolio Manager** benchmarks energy and water performance but is not used for LEED certification submissions. **EDUCATION @USGBC** is the education portal for courses and credentials.

**60) Correct answer:** C. The climate risk.

While a hazard describes what could happen (flooding, extreme heat), climate risk refers to the potential adverse consequences of that hazard actually occurring — including impacts on the building, its occupants, and the owner's investment. The team has already completed the hazard assessment by identifying flooding and extreme heat. Now they are evaluating climate risk by estimating the potential consequences of a flood on mechanical systems and occupant displacement. Social determinants of health address living conditions.

**61) Correct answer:** C. In the owner's project requirements (OPR).

LEED requires that specific, measurable project goals addressing the impact areas be incorporated into the owner's project requirements (OPR) first. The OPR is the foundational document that captures the owner's vision and performance requirements, and all subsequent design decisions flow from it. The commissioning provider's report verifies performance against the OPR but does not originate goals. The LEED scorecard tracks credit selection. The waste management plan addresses construction waste, not operational carbon targets.

**62) Correct answer:** A. Greenwashing.

**Greenwashing** is the presentation of a product or a material as being more environmentally friendly than it actually is. **Green chemistry** is the design of chemical products and processes that eliminate the use and generation of hazardous substances. Green misleading and green marketing are not LEED terms.

**63) Correct answer:** B. Source reduction.

Source reduction refers to the exact sizing of the materials to be produced through prefabrication, modular construction, or similar methods, in order to prevent waste. Since waste is not generated at all, source reduction is the most effective strategy to reduce solid waste.

EPA developed a hierarchy, ranking the most environmentally sound strategies for reducing municipal solid waste:

1. Source reduction
2. Reuse
3. Recycle
4. Waste-to-energy conversion

**64) Correct answers:** A. Environmental stewardship. B. Social equity and justice. D. Economic prosperity.

All the green buildings should establish three components:

- Economic prosperity – (Profit)
- Social equity and justice – (People)
- Environmental stewardship – (Planet)

Those 3 components are called the triple bottom line, also referred to as the three pillars of sustainability. A commitment to the triple bottom line means looking beyond a project's profitability, or considering more than just a project's sustainable features, or social responsibility.

**65) Correct answers:** A. Native plants. B. Xeriscaping. E. Adapted plants.

Using native and/or adapted plants and xeriscaping reduces outdoor water use.

**Native plants (indigenous plants)** are the type of plants that occur and develop naturally in a specific location, and can thrive without much maintenance. Therefore, using native plants will make a big contribution to the outdoor water use reduction.

**Xeriscaping** is a type of landscaping that does not need any irrigation.

**Invasive plants** that exist on-site should always be cleaned, which will in turn take over the adjacent existing native and adapted plants. Invasive plants are the types of plants that spread and damage the environment. For this question, they will not make any positive contribution to the outdoor water use.

Areas covered with **turfgrass** will need high amounts of irrigation water.

**Adapted plants** are the types of plants that do not occur naturally in a specific location, however, they can nonetheless adapt easily to the climate of the region. Adapted plants and native plants can both thrive without extensive irrigation, pesticides, and fertilizers.

**66) Correct answer:** A. Project Priorities credit.

Project Priorities credit awards up to 9 points for pursuing any combination of the following pathways in the Project Priority Library: Exemplary Performance, Regional Priority, Project-Type Credits, Pilot Credits, and Innovative Strategies.

**67) Correct answer:** B. Efficiency of the filtration media.

Minimum efficiency reporting value (MERV) rates the efficiency of the filtration media.

**68) Correct answer:** B. Social determinants of health.

**Social determinants of health** are the conditions in which people are born, live, and work that shape well-being — including access to clean air, safe housing, green spaces, nutritious food, and quality education. A community lacking these resources faces health disparities driven by these underlying conditions. Addressing social determinants goes beyond the building itself to consider the community's broader needs. **Resilience** addresses a building's ability to withstand disruptions. **Passive survivability** refers to maintaining life-support conditions during outages. **Demand response** relates to electricity management.

**69) Correct answer:** D. Any action taken in a leverage point of the system will have no effect.

A small change can yield big changes. In other words, any action taken in a **leverage point** of the system can bring about significant results in the system. Think about a construction site that is located close to a river, and that river is the main irrigation source of the surrounding farmlands. If the construction site does not take any action to stop sedimentation and pollution resulting from the construction activities, the whole river can then become contaminated. This will then contaminate the farmlands, which will result in contaminated foods. However, by taking action to avoid sedimentation and pollution to the river, which is really a small action when considering the outcome, a whole river can be saved that will also result in saving the farmlands and human health. **Therefore, any action taken in a leverage point of the system will have the biggest effect.**

A **positive feedback loop** can be summarized as A producing B, which in turn produces more of A. Therefore, an interest savings account can be an example of a positive feedback loop, as the account grows, more interest is earned which in turn brings further account growth.

**Open systems** are systems that constantly consume other items, use them, and produce waste at the end. A whole city may be an example of this, as it imports food, energy, and products; however, while a city uses products, it also produces wastes in turn, often in the form of sewage, pollution, and solid waste, which consequently marks the end of the system. There are no **feedback loops** in the open systems.

However, in nature, everything is in a loop, and thus no open system exists. Therefore, this is called a **closed system** that does not produce any waste product at the end. The system continues independently. Think about the water cycle in our world: the sun heats and evaporates the water on the oceans and the seas, and then creates water vapor. The resulting water vapor rises on air, which condenses into rain and goes back to the earth's surface. That water then becomes underground water, which again joins the oceans and seas with the flow. This sustainable cycle does not end.

**70) Correct answer:** B. Untreated marble.

Naturally occurring materials and products made from inorganic materials such as marble or granite contain low or no VOCs and are therefore exempt from LEED requirements. These materials are called **inherently nonemitting materials**.

**71) Correct answer:** A. Water from toilets.

Water from toilets would be classified as **blackwater**, which is the term to describe the used water that has come into contact with waste.

**Graywater**, on the other hand, is the untreated household water that did not come into contact with toilet waste. Used water from bathtubs, showers, bathroom wash basins, and water from clothes washers and laundry tubs can be examples of graywater, and may be used as a flush water in toilets or urinals. However, this definition can change depending on the local codes, therefore projects should first check with local codes before using graywater, and then meet the LEED requirements.

**72) Correct answer:** B. LEED Residential.

Only LEED Residential projects require in-field verification.

**73) Correct answers:** C. Appeal fees. D. Credit interpretation ruling (CIR).

D. Credit interpretation ruling (CIR) fees.

Appeal and credit interpretation ruling (CIR) fees are flat fees, and are the same for both USGBC members and non-members.

Review fees depend on the <u>gross floor area of the project</u>, and the <u>membership status</u>. USGBC member companies pay less for the review fees.

Project registration fees do not depend on the size of the project, but depend on the <u>membership status</u>, and USGBC member companies pay less for the project registration fees.

**74) Correct answer:** B. U.S. Endangered Species Act and the IUCN Red List.

The U.S. Endangered Species Act defines endangered and threatened species at the federal level, while the International Union for Conservation of Nature (IUCN) Red List identifies endangered and threatened species globally. NatureServe also assigns conservation status ranks but was paired with an unrelated tool in option D.

**75) Correct answer:** C. Sufficiency.

**Sufficiency** refers to measures and practices that avoid unnecessary demand for energy, materials, land, and water while still delivering human well-being within planetary boundaries. It is the broadest of the material-related concepts — encompassing the philosophy of "enough, not more."

**Dematerialization** specifically reduces material quantities in a building.

**Material circularity** keeps materials in use through reuse and recycling.

**Adaptive reuse** designs buildings for future flexibility.

Sufficiency is the overarching mindset that drives all of these individual strategies.

**76) Correct answer:** A. Pre-design phase.

To establish an integrated building process, a working system needs to be created. And the system should be active as early as possible during the **pre-design phase** or even before deciding on the project location. Think about the following questions:

- How is the climate in the project's location, what type of heating and cooling system should be installed?
- How should the building be oriented in order to get more sunlight into the building, and to reduce the operating costs and provide natural daylighting?
- Is there enough rain fall on the location to capture and use rainwaterfor irrigation?
- What are the goals and requirements of the project's owner?
- How will the building occupants commute once the project is finished, what are the public transportation options?
- How will the resources and construction materials will be transported to the project site?
- What are the building code requirements in the project's location?
- How does energy get to the site?
- What is the type of soil on-site?
- What type of vegetation exists on-site?

They're really the type of questions that should be addressed during the earliest phase of the project. This is why, the key in the integrated process is to integrate the <u>whole</u> project team early in order to develop an efficient project design, and allow the project team to discover strategies. Taking actions late in the process will only result in delays and additional costs.

**77) Correct answer:** B. Facility manager.

Facility manager is the person that will be responsible for running the building operations after the completion of construction.

**78) Correct answers:** A. Third-party consultant with no involvement in the project's design or construction. B. Owner's employee. E. Completely independent employee of the design or construction firm.

The CxP can be a third-party consultant with no involvement in the project's design or construction, an employee of the owner provided they meet the qualification requirements, or an employee of the design or construction firm as long as they are completely independent from any aspect of the project's design or installation. A contractor's or subcontractor's employee cannot qualify without that independence, because they are directly associated with the systems being commissioned. Where any potential conflict of interest exists, the commissioning plan must demonstrate that the CxP acts solely on behalf of the owner and reports directly to the owner.

**79) Correct answer:** D. Owner.

In every case the commissioning provider (CxP) must directly report to the owner.

**80) Correct answer:** C. Regenerative design does not contribute to the restoration of the lost environment and habitat.

**Regenerative design** can be seen to be one step ahead of sustainable design, which is a type of building design that creates no waste, and provides more output than consumed input. As regenerative design projects do not use any more energy than they produce, they are able to return the excess energy to the utility. In addition to their sustainable features, they can also contribute to the restoration of the lost environment and habitat, thereby taking regenerative design one step ahead of sustainable design.

**81) Correct answer:** B. Reducing energy use through better design or energy retrofits.

The main decarbonization actions follow a sequence: first reduce energy use through better design or energy retrofits, then electrify building systems and optimize how the project interacts with the power grid, then maintain low energy use and emissions over time, and finally produce and procure clean energy. Efficiency comes first because the cleanest and cheapest energy is the energy never used — a leaner building needs smaller systems and less renewable generation to reach the same goal. Offsets and procurement address what remains; they are not a substitute for reducing demand.

**82) Correct answer:** B. Systems thinking.

**Systems thinking** refers to the understanding of each and every system of a building, while also understanding their relationships and looking at the project as a whole. **Feedback loop** is system structure that causes output from one node to eventually influence input to that node. **Leverage points** are places within a complex system (a corporation, an economy, a living body, a city, an ecosystem) where a small shift in one thing can produce big changes in everything.

In the conventional building approach, project team member completes a work individually and then passes it to the next person, at this is called the **linear approach**. However, in the integrated process, linear approach does not exist, coordination of all the team members is required.

**83) Correct answer:** C. U.S. EPA 2022 Construction General Permit.

LEED requires all projects to follow the erosion and sedimentation requirements of the U.S. EPA 2022 Construction General Permit (CGP), the EU Taxonomy DNSH provisions, or a local equivalent. ASHRAE 62.1 addresses ventilation. The ITE Parking Generation Manual addresses parking capacity. The ANSI Guide 2.0 addresses the integrative process.

**84) Correct answer:** A. 1,000 square feet (93 square meters) gross floor area.

According to LEED Minimum Program Requirements (MPRs), projects must comply with specific size thresholds to be eligible for certification. LEED BD+C and LEED O+M projects must include a minimum of 1,000 square feet (93 square meters) of gross floor area. LEED ID+C projects must include a minimum of 250 square feet (22 square meters) of gross floor area.

In addition to meeting size requirements, all LEED projects must satisfy the other two Minimum Program Requirements: the project must be in a permanent location on existing land and must use reasonable LEED boundaries.

**85) Correct answers:** A. Project registration fees. B. Project review fees.

Review fees depend on the gross floor area of the project, and the membership status. USGBC member companies pay less for the review fees.

Project registration fees do not depend on the size of the project, but depend on the membership status, and USGBC member companies pay less for the project registration fees.

However fees for LEED interpretations do not depend on the membership status.

LEED Prototype fees do not exist.

**86) Correct answer:** D. Positive feedback loop.

A positive feedback loop occurs when a change produces more change in the same direction — A produces B, and B produces more of A. Development attracts businesses, businesses attract residents, and residents drive further development — each element reinforces the others in an escalating cycle. A negative feedback loop restores balance by producing an opposite response. A leverage point is where a single action creates significant results. Emergent properties arise from system interactions.

**87) Correct answer:** C. Biophilic design.

**Biophilic design** integrates nature into the built environment through direct contact with nature (plants, water, daylight, fresh air, views), natural materials and forms (wood, stone, organic textures and patterns), and nature-inspired spatial layouts. Live plants, natural wood, and organic patterns are all biophilic elements. **Regenerative design** aims to give back more than consumed. **Inclusive design** makes buildings accessible to all abilities. **Adaptive reuse** accommodates future layout changes.

**88) Correct answer:** C. 100% of special-status vegetation as defined by local, state, or federal entities.

LEED requires all projects to protect 100% of special-status vegetation on-site, as defined by local, state, or federal entities. This does not mean all existing vegetation must be preserved — only species that carry a special designation due to their ecological importance, rarity, or protected status. The IUCN Red List identifies endangered species globally but is not the sole reference for on-site vegetation protection. Native plants are valued by LEED but are not all automatically classified as special-status.

**89) Correct answer:** A. U.S. Department of Agriculture U.S. Code of Federal Regulations Title 7, Volume 6.

US Department of Agriculture, United States Code of Federal Regulations Title 7, Volume 6—Defines the prime farmlands.

US Environmental Protection Agency (EPA)'s National Priorities List—Defines national priority sites.

US Housing and Urban Development—Defines Federal Empowerment Zone, Federal Enterprise Community, and Federal Renewal Community.

US Department of Treasury, Community Development Financial Institutions Fund—Provides funds for low-income communities.

**90) Correct answer:** D. Water from the public utilities.

Alternative water sources are non-potable water from on-site surfaces, or freshwater sources, such as graywater, on-site reclaimed water, collected rainwater, captured condensate, and rejected water from reverse osmosis systems. Water from the public utilities cannot be considered as an alternative water source.

**91) Correct answer:** B. High-angle front lights.

Glare occurs when a light source within a person's field of view produces excessive brightness or contrast, causing visual discomfort and reducing the ability to see clearly. High-angle front lights — fixtures positioned in front of and above the occupant — direct light toward the eyes at an angle that falls within the typical line of sight, creating direct glare.

**Backlight** is the primary factor of **light trespass** onto the adjacent sites, and **uplight** is the cause of artificial glow.

**92) Correct answer:** B. Green Raters.

**LEED Green Raters** provide in-field verification to every LEED Residential project. (The other rating systems do not require any in-field verification.)

The **Green Classroom Professional (GCP)** certificate is designed for educators and school facility professionals who want to gain a foundational understanding of green building principles in the context of a school setting, helping them identify what supports or impedes healthy, resource-efficient, and environmentally sustainable learning spaces.

**LEED Green Associate** credential is for professionals who want to demonstrate green building knowledge and expertise in nontechnical fields of practice, to denote basic knowledge of green design, construction and operations. LEED Green Associates have a proven, up-to-date understanding of green building principles and practices.

**93) Correct answer:** D. Space volume.

Thermal comfort is one of the top priority measures to increase occupant comfort, which is defined by a combination of six factors: air temperature, humidity, surface temperature, air movement, metabolic rate, and clothing.

**94) Correct answer:** D. Appeal.

Phases of the LEED certification process are:

- Project Registration
- Application
- Review
- Certification or Denial

**95) Correct answer:** A. Dry ponds.

**Dry ponds** hold the excess rainwater for some time, thereby allowing the rainwater to slowly seep into the ground without contamination. **Bioswale** is a stormwater control feature which uses a combination of engineered basin, soils and vegetation. **Low-impact development (LID)** is an approach to mimic natural systems and to manage the stormwater as closest to its source. **Green infrastructure (GI)** is named for the infrastructure to direct the rainwater collected from the impervious surfaces to the vegetation and soil surfaces without giving them to the storm sewer system.

**96) Correct answer:** C. Spray irrigation.

Spray irrigation systems are conventional systems and are not water efficient. High-performance irrigation systems are drip irrigation systems (most water efficient system), bubbler distribution systems, and channel water directly to root systems.

**97) Correct answer:** C. Low-impact development (LID).

**Low-impact development** is a rainwater management strategy, which is an approach to mimic natural systems and to manage the stormwater as closest to its source. LID strategies include decreasing impervious surfaces and increasing vegetation on-site, and in addition, LID works on slowing down the flow of rainwater out of site and makes sure the rainwater does not get contaminated. And for this question, LID is not relevant to reducing greenhouse gases.

Utilizing a **compressed workweek** can be a strategy to reduce fossil fuel usage. A compressed workweek refers to when companies allow employees to work more hours during working days in exchange for a day off. Alternatively, companies could utilize **telecommuting** for a part of the working week, so that the employers can work from their homes without commuting to the office. Carpooling can also be rewarded, while the parking rates can be made more expensive to discourage individual vehicle use.

**Electric vehicles** do not rely on burning fossil fuels and make a significant contribution to reducing greenhouse gas emissions.

**98) Correct answer:** A. Solar reflectance index (SRI).

**Solar reflectance (SR)** value shows the solar energy that is reflected by a surface on a scale of "0" to "1". A black surface will have a SR of "0", while a white surface will have a SR of "1". SR does not does not consider the emissivity of the material.

**Albedo** is a type of reflectivity measurement from "0" to "1", which "0" represents black surfaces that absorb all the solar radiation, while "1" represents white surfaces that reflects all the solar radiation. Albedo does not does not consider the emissivity of the material.

**Heat island effect** is not a reflectivity measurement. Dark colored, non-reflective surfaces absorb heat during hot weather, and release it into the atmosphere, and this releasing of heat is called the heat island effect.

**99) Correct answer:** C. Heat island effect.

**Albedo** is a type of reflectivity measurement from “0” to “1”, which “0” represents black surfaces that absorb all the solar radiation, while “1” represents white surfaces that reflects all the solar radiation. Albedo does not does not consider the emissivity of the material. **Emissivity** (**infrared** or **thermal emittance**) is a measure which shows how much heat or infrared radiation a material can shed back into the atmosphere.

**100) Correct answer:** D. Encourage light pollution reduction.

Encouraging light pollution reduction is the goal of the Sustainable Sites credit category.

# SECTION 3:

## PRACTICE TEST 2

**1)** For some of the Water Efficiency prerequisites/credits, projects that contain landscaping should calculate the baseline theoretical irrigation requirement (TIR). TIR is the amount of irrigation water that the landscape will require:

a) During the site's peak watering month
b) During the site's average watering month
c) During the summer months only
d) Annually

**2)** Which of the following waste management strategies is not mandatory for LEED projects?

a) Safe collection, storage, and disposal of e-wastes generated by the building occupants
b) Recycling of mixed paper, corrugated cardboard, glass, plastics, and metals produced by the building occupants
c) Providing dedicated areas accessible to waste haulers and building occupants
d) Sorting construction materials into streams for recycling

**3)** A project team installs submeters on all major water-consuming systems and uses the data to optimize cooling tower operations. Which credit categories benefit from this strategy?

a) Water Efficiency only
b) Water Efficiency and Energy and Atmosphere
c) Energy and Atmosphere only
d) Sustainable Sites and Water Efficiency

**4)** Which of the following are the phases of the split review option for the LEED BD+C and LEED ID+C project certification?

a) Construction
b) Anticipated and awarded
c) Design and construction
d) Preliminary review and final review

**5)** Which of the following is not an outcome of the compact development strategies?

a) Reduction in the greenhouse gas emissions
b) Increased availability of alternative transportation
c) Increased need for infrastructure and roads
d) Development in dense areas

**6)** Which of the following spaces does not qualify as an open space under the LEED requirements?

a) Pedestrian oriented paving or turf area with physical site elements that accommodate outdoor social activities for building occupants
b) Garden space with different vegetation types that is visible to the building occupants but not physically accessible
c) Garden space dedicated to community gardens or urban food production
d) Recreation oriented paving or turf area with physical site elements that encourage physical activity for building occupants

**7)** Which of the following is the minimum level of LEED certification a project can receive?

a) Bronze
b) Silver
c) Gold
d) Certified

**8)** In a hot-arid climate, which scenario would most improve a project's energy performance?

a) Increasing ventilation rates by 30%
b) Installing ENERGY STAR appliances only
c) Improving envelope insulation and HVAC efficiency
d) Using lighting fixtures with LED bulbs

**9)** Which of the following would assist in reducing single-occupancy vehicle use? (Choose 2)

a) Limiting the number of parking spaces
b) Increasing parking rates
c) Locating the project further away from alternative transportation
d) Reducing the number of bicycle parking spaces

**10)** Which of the following is the biggest contributor to global climate change?

a) Increase in the carbon dioxide levels (CO2)
b) Construction wastes
c) Heat island effect
d) Development of greenfields

**11)** Which of the following is not an alternative fuel?

a) Hydrogen
b) Propane
c) Electricity
d) Diesel

**12)** Which of the following specifies the green building products to be used (such as products that meet Green Seal, Environmental Choice, or EPA standards), chemicals allowed to be used inside the building, training of the cleaning personnel for the use of chemicals and green cleaning practices, indoor pest control plans, and energy efficient cleaning equipment?

a) Green purchasing
b) Green cleaning program
c) Sustainable material purchasing program
d) Green chemistry

**13)** Which of the following is not a soft cost?

a) Supervision costs
b) Design cost
c) Labor cost
d) Permits and taxes

**14)** "A producing B, which in turn produces more of A" can be an example of which of the following?

a) Negative feedback loop
b) Positive feedback loop
c) Open system
d) Emergent property

**15)** A project team is pursuing LEED BD+C certification. The building has solar panels generating renewable energy on-site. The project team retains the renewable energy certificates (RECs). A neighboring building owner asks to purchase the RECs. If the project team sells the RECs, what happens to the project's renewable energy claim under LEED?

a) The claim is unaffected because the electricity is still physically consumed on-site

b) The project loses the ability to claim on-site renewable energy under LEED because the environmental attributes have been transferred

c) The project can still claim 50% of the renewable energy since the panels remain on-site

d) The project must purchase carbon offsets to replace the sold RECs

**16)** Which of the following describes the information necessary to accomplish the owner's project requirements, and also includes system requirements, project design criteria, standards, and guidelines?

a) Commissioning (Cx) plan

b) Basis of design (BOD)

c) Cutsheets

d) Ordinances

**17)** What is the minimum outdoor water use reduction from the baseline theoretical irrigation requirement (TIR) that a project with a permanent irrigation system must achieve under the LEED BD+C: New Construction and Major Renovations rating system?

a) 10%

b) 20%

c) 30%

d) 40%

**18)** At the end of the construction review of a LEED BD+C project, GBCI will mark the submitted prerequisites and credits as: (Choose 3)

a) Anticipated

b) Awarded

c) Denied

d) Clarify

e) Earned

**19)** Which of the following is not a rainwater management strategy?

a) Use of bioswales and vegetated filters

b) Low-impact development (LID)

c) Green infrastructure (GI)

d) Efficiently directing runoff into the municipal storm sewer

**20)** Which of the following has the lowest solar reflectance (SR) value?

a) Black asphalt surface
b) White marble surface
c) Light-colored roof
d) Soil surface

**21)** A LEED project team registered their project in June 2026. In April 2026, USGBC published a LEED interpretation that changes how a prerequisite the team is pursuing should be documented. The team was unaware of this interpretation and has been following the original reference guide approach. What is the team's obligation?

a) The team has no obligation because they were unaware of the interpretation at the time of registration
b) The team should follow the reference guide because it was published before the interpretation
c) The team must comply with the interpretation because it was published before their project registration date
d) The team should request a waiver from GBCI since they were already following the reference guide

**22)** A project team is evaluating the embodied carbon of their building materials. Which of the following materials would typically have the highest embodied carbon?

a) Locally sourced, FSC-certified timber framing
b) Recycled steel structural members
c) Conventional Portland cement concrete
d) Rapidly renewable bamboo flooring

**23)** What does vehicle miles traveled (VMT) measure in the context of LEED?

a) The total distance construction materials are transported to the project site
b) The estimated annual miles driven by building occupants commuting to and from the project
c) The total distance waste haulers travel to remove waste produced by the building occupants
d) The cumulative mileage of all vehicles parked on the project site

**24)** The three pillars of sustainability are embedded in which international framework of 17 goals that green building helps advance?

a) The Paris Agreement
b) The United Nations Sustainable Development Goals (SDGs)
c) The Montreal Protocol
d) The GHG Protocol

**25)** Which of the following has the biggest effect on a building's energy demand?

a) Building orientation
b) Configuration of the adjacent buildings
c) Community planning
d) Building height

**26)** Why does reducing a building's potable water use also reduce carbon emissions beyond the project site?

a) Treating and pumping water consumes large amounts of energy, often fossil-fueled
b) Water itself releases carbon dioxide as it evaporates
c) Lower water use directly reduces the building's refrigerant charge
d) Water bills fund fossil fuel extraction

**27)** In LEED O+M, building performance data must be collected over a defined timeframe before certification can be submitted. What is this timeframe called?

a) Commissioning period
b) Performance period
c) Monitoring period
d) Evaluation cycle

**28)** Which of the following should the project team consider first after deciding to use graywater as an alternative to potable water in irrigation?

a) LEED requirements
b) Local code requirements
c) WaterSense requirements
d) Americans with Disabilities Act (ADA)

**29)** Which of the following are types of plants that occur and develop naturally in a specific location? (Choose 2)

a) Indigenous plants
b) Adapted plants
c) Native plants
d) Invasive plants

**30)** When choosing environmentally friendly products, a project team should look for: (Choose 3)

a) Health product declarations (HPD)
b) Environmental product declarations (EPD)
c) Corporate sustainability reports (CSR)
d) Custodial effectiveness assessments
e) Greenwashing

**31)** Which of the following refers to material that is recycled before being used by a consumer?

a) Preconsumer recycled content
b) Postconsumer recycled content
c) Extended Producer Responsibility (EPR)
d) Total recycled content

**32)** Which of the following certifications / organizations does not address indoor air quality by setting requirements for the VOC levels of the materials used?

a) South Coast Air Quality Management District (SCAQMD)

b) Green Label (Carpet and Rug Institute)

c) FloorScore

d) Green-e

**33)** Which of the following statements is false regarding inherently non-emitting materials?

a) LEED requires projects not to use inherently non-emitting materials

b) Inherently non-emitting materials refers to the VOC level of a material

c) Inherently non-emitting materials are products without any VOC emissions

d) Brick, concrete, and ceramic can be some examples of inherently non-emitting materials

**34)** Which of the following is not an entryway system?

a) Grilles

b) Grates

c) Filtration media

d) Rollout mats

**35)** Which principle holds that all people should have equal protection from environmental hazards and equal participation in decisions that affect their environment?

a) Social determinants of health

b) Inclusive design

c) Environmental justice

d) Sustainable development

**36)** Which of the following statements is false if a project team decides to create their own Innovation credit?

a) The proposed credit should demonstrate measurable environmental performance

b) The proposed credit should be significantly better than then the standard sustainable design practices

c) The proposed credit shoud not be addressed in the LEED rating systems

d) The proposed credit should be applicable to all the LEED rating systems

**37)** A building requires high outdoor air ventilation rates due to its occupancy type, which significantly increases heating and cooling loads. Which strategy would best reduce the energy impact of conditioning this outdoor air?

a) Demand-controlled ventilation

b) Variable air volume (VAV) system

c) Energy recovery ventilation (ERV)

d) Increasing duct insulation

**38)** A project cannot earn any exemplary performance points by achieving:

a) Minimum Energy Efficiency prerequisite
b) Electrification credit
c) Renewable Energy credit
d) Construction and Demolition Waste Diversion credit

**39)** The use of which of the following, will not result in increased daylighting?

a) Light shelves
b) Clerestory windows
c) Low height cubicle partitions
d) Glare control devices

**40)** During a power outage, a building's on-site solar panel system automatically disconnects from the utility grid and continues supplying electricity to critical building systems. What is this capability called?

a) Load shedding
b) Demand response
c) Grid interactivity
d) Islanding

**41)** Which of the following is the former blackwater that has been treated and purified for reuse, and can even be used for irrigation?

a) Graywater
b) Reclaimed water
c) Black water
d) Potable water

**42)** Which of the following is false about the LEED green building rating system?

a) LEED rating systems provide third party verification
b) LEED rating systems are consensus based rating systems
c) LEED rating systems are developed by a profit organization
d) LEED rating systems have a fair and open review process

**43)** A building's energy use intensity (EUI) is significantly higher than comparable buildings in the same climate zone. Which professional should the facility manager engage to systematically investigate and optimize the building's performance?

a) A commissioning provider (CxP)
b) A monitoring-based commissioning provider (MBCxP)
c) A LEED Project Administrator
d) An energy rater

**44)** Which of the following cannot be evaluated during the site assessment?

a) Hydrology
b) Nonpoint source pollution
c) Site topography
d) Human use

**45)** A new building will have 42% of its gross floor area without finished interiors at the time of certification. Which rating system must the project use?

a) LEED BD+C: New Construction, because more than half the building is complete
b) LEED O+M: Existing Buildings, because the incomplete spaces will be finished later
c) LEED ID+C: Commercial Interiors, because the incomplete spaces need interior fit-out
d) LEED BD+C: Core and Shell, because at least 40% of the floor area is incomplete

**46)** Which of the following is not a strategy directly related with reducing lighting costs?

a) Using reflective paving materials to distribute the light across the site
b) U.S. Department of Energy's Commercial Buildings Energy Consumption Survey (CBECS)
c) Installing occupancy sensors
d) Installing individual lighting controls

**47)** How much would be the gross floor area of a building, if it has a floor-to-area ratio (FAR) of 3, and a total lot area of 10,000 square feet?

a) 15,000 sf
b) 10,000 sf
c) 30,000 sf
d) 3,333 sf

**48)** For all LEED BD+C: New Construction and Major Renovations projects, at least ________ of the project's gross floor area must be complete at the time of LEED certification.

a) 40%
b) 50%
c) 60%
d) 70%

**49)** Which of the following tasks cannot be performed using Arc?

a) Registering the project for the LEED certification
b) Submit LEED application to GBCI for certification review
c) Complete prerequisite and credit documentation
d) Searching for Addendas

**50)** During which phase of the integrated process are life-cycle costing, life-cycle assessment, and computer modeling used to evaluate different strategies and alternatives?

a) Occupancy, operations, and performance feedback phase
b) Design and construction phase
c) Discovery phase
d) Commissioning phase

**51)** Using open-grid paving system as a part of the hardscape can aid in earning which of the following credits? (Choose 2)

a) Minimum Water Efficiency
b) Heat Island Reduction
c) Rainwater Management
d) Transportation Demand Management

**52)** Which of the following can decrease rainwater runoff by replicating natural hydrology and help re-create the water balance of the project site? (Choose 2)

a) Green infrastructure (GI)
b) Hardscape areas
c) Efficient storm drains on-site
d) Low-impact development (LID)

**53)** LEED projects should establish a recycling program that collects which of the following materials? (Choose 3)

a) Metals
b) Glass
c) Wood products
d) Mixed-paper
e) Batteries

**54)** Which of the following is not a LEED impact area?

a) Ecological Conservation and Restoration
b) Social Equity and Justice
c) Quality of Life
d) Decarbonization

**55)** A school district wants its facility staff to better understand how green building principles apply specifically to educational environments. Which USGBC credential is designed for this purpose?

a) LEED Green Associate
b) Green Rater
c) Green Classroom Professional (GCP) certificate
d) LEED Fellow

**56)** A LEED Project Administrator should have which of the following accreditations?

a) LEED Fellow

b) LEED Green Associate

c) LEED AP with specialty

d) Does not need to have an accreditation

**57)** Which of the following developed the WaterSense label?

a) U.S. Green Building Council

b) U.S. Environmental Protection Agency

c) EnergySTAR

d) Green Seal

**58)** Project teams can earn a point by employing which of the following as a principal participant of the project?

a) LEED Legacy AP

b) LEED AP BD+C

c) LEED Green Associate

d) LEED Project Administrator

**59)** A community adjacent to an industrial facility experiences disproportionately high rates of asthma and respiratory illness. A new LEED project is proposed nearby. Which LEED principle should the project team consider to address this community's needs?

a) Material circularity

b) Compact development

c) Environmental justice

d) Dematerialization

e) GBCI

**60)** In order to achieve the Rainwater Management and Accessible Outdoor Space credits, a project team decides to reduce the building footprint and increase greenfield area. However, local codes do not allow the reduction of the building footprint. What should the project team do?

a) Submit a credit interpretation ruling (CIR) and ask for a clarification

b) Inform GBCI about the local codes, and also provide documentation to achieve the credits without reducing the building footprint

c) Reduce the building footprint and inform the local authorities about the reasons for not obeying the local codes

d) Follow the local codes instead of LEED requirements

**61)** Which of the following best describes the primary purpose of an erosion and sedimentation control (ESC) plan during construction?

a) To prevent stormwater runoff from leaving the construction site entirely

b) To prevent soil loss, protect waterways from sedimentation, and reduce airborne dust

c) To restore disturbed soils and revegetate the site after construction is complete

d) To manage construction waste and divert materials from landfill

**62)** Which of the following are addressed by Minimum Program Requirements (MPRs)? (Choose 2)
- a) Project boundaries
- b) Project's size
- c) Building orientation
- d) Building performance

**63)** ASHRAE Standard 55-2023 is used for:
- a) Energy reductions
- b) Indoor air quality
- c) Thermal comfort
- d) Building energy efficiency

**64)** Which of the following is not a LEED BD+C credit category?
- a) Water Efficiency
- b) Location and Transportation
- c) Project Priorities
- d) Energy and Resources

**65)** Which of the following rating systems can the tenants of a LEED BD+C: Core and Shell office building additionally pursue for their office space? (Choose 2)
- a) LEED O+M: Core and Shell
- b) LEED BD+C: Offices
- c) LEED ID+C: Commercial Interiors
- d) LEED O+M: Existing Buildings

**66)** Buildings certified under the LEED for Interior Design and Construction (ID+C) rating systems should pursue recertification every:
- a) 2 years
- b) 3 years
- c) 5 years
- d) LEED ID+C projects do not require recertification

**67)** Which of the following is a previously developed site that was contaminated with waste or pollution?
- a) High-priority site
- b) Brownfield site
- c) Sensitive land
- d) A Federal Empowerment Zone site

**68)** Under the LEED requirements, the erosion and sedimentation control (ESC) plan of all the newly built projects should conform to which of the following?

a) U.S. EPA 2022 Construction General Permit (CGP)

b) EPA Level III Ecoregion requirements or local equivalent

c) Environmental Protection Agency (EPA) water budget requirements or local equivalent

d) ASHRAE Guideline 0-2005 or local equivalent

**69)** Which of the following refers to SMACNA Guidelines?

a) Minimum energy performance

b) Renewable energy production

c) Construction indoor air quality management plan

d) Quality views

**70)** A project team is designing a new office building and wants to maximize energy performance through an integrative approach. Which of the following strategies would most effectively reduce the building's overall energy consumption?

a) Installing a high-efficiency HVAC system sized for the building's original design loads

b) Optimizing the building envelope and passive design strategies first, then sizing mechanical systems based on the reduced loads

c) Purchasing renewable energy credits (RECs) to offset the building's projected energy use

d) Specifying ENERGY STAR-rated equipment and appliances throughout the building

**71)** A building is designed to maintain safe indoor temperatures and basic lighting for 72 hours during a complete power outage through passive design strategies and battery storage. Which concept does this capability demonstrate?

a) Demand response

b) Grid interactivity

c) Resilience

d) Load shifting

**72)** In the LEED point allocation process, points are allocated to each credit according to a credit's contribution to the:

a) LEED guiding principles

b) LEED impact areas

c) LEED credit categories

d) LEED priority list

**73)** Which of the following are natural refrigerants? (Choose 3)

a) Ammonia

b) Helium

c) Carbon dioxide

d) CFC-11

e) Propane

**74)** Which of the following is not a sensitive land?

a) Prime farmland

b) Brownfield site

c) Sites below floodplain

d) Wetlands and waterbodies

**75)** A project team has completed the human impact assessment and identified that the project site lacks pedestrian sidewalks and has no bus stops within walking distance. Under which assessment category were these findings identified?

a) Human use and health impacts

b) Occupant experience

c) Demographics

d) Local infrastructure and land use

**76)** Which of the following is for organizations planning to certify a large number of new construction buildings or interior spaces based on a common prototype?

a) LEED Campus Program

b) LEED Volume Program

c) LEED Prototype Program

d) LEED Uniform Program

**77)** When applying for the Project Priorities credit, a project team:

a) Can receive points for each of the LEED Accredited Professionals on the team

b) Can achieve a former pilot credit and receive point(s), even if the pilot credit is not available on the USGBC's pilot credit library at the time the project decides to register for it

c) Can achieve exemplary performance points for the qualifying prerequisites

d) Receive points by achieving significant, measurable environmental performance using an innovative strategy which is not addressed in the LEED green building rating system

**78)** In LEED, which of the following cannot be counted as materials diverted from landfills for the construction waste management? (Choose 3)

a) Land clearing debris

b) Alternative daily cover

c) Petroleum based plastics

d) Wood based products

e) Hazardous materials

**79)** Using air filters with high MERV values will result in: (Choose 2)

a) Increased energy usage

b) Increased removal of particles and contaminants from the supply air

c) Decreased air quality

d) Decreased thermal quality

**80)** Which of the following is not a renewable energy source?

a) Wind
b) Wave
c) Biomass
d) Natural gas

**81)** Why are heat pumps the most efficient systems available for space heating and water heating?

a) They burn fuel at higher temperatures than boilers
b) They move existing thermal energy from one place to another rather than generating heat
c) They run only during off-peak hours
d) They eliminate the need for electricity

**82)** Which of the following enables international projects to earn the appropriate prerequisites / credits by allowing them to meet international standards or their local standards, instead of U.S. based standards?

a) Alternative compliance paths (ACPs)
b) Addendas
c) Credit Interpretation Rulings (CIR)
d) Alternative credits

**83)** Which of the following items does not consume process energy (unregulated energy)?

a) Kitchen stoves
b) Office equipment
c) Elevators and escalators
d) HVAC equipment

**84)** Which of the following statements is false?

a) LEED interpretations are administered by GBCI
b) LEED interpretations are subject to a USGBC consensus based review
c) Adherence to LEED interpretations posted after a project registers is optional but encouraged
d) Projects are required to adhere to LEED interpretations posted before their registration date

**85)** Which of the following is not a strategy to reduce the indoor potable water consumption of a building?

a) Installing High-Efficiency Toilets (HETs)
b) Installing waterless urinals
c) Installing composting toilet systems
d) Installing drip-irrigation systems

**86)** A project team has completed construction of a new building and is preparing to occupy the space. The owner wants to ensure that all energy-related systems perform as intended and continue to operate efficiently over time. Which of the following actions would best support this goal?

a) Conducting a one-time energy audit at the time of occupancy

b) Engaging a commissioning provider to verify system performance during design, construction, and post-occupancy

c) Installing advanced energy meters

d) Requiring the contractor to provide warranties on all installed equipment

**87)** In LEED, which of the following are not considered qualifying use types that count toward a project's proximity to community services and amenities? (Choose 2)

a) ATMs

b) Restaurants

c) Bank

d) Vending machines

**88)** Which of the following are conducted among the project team, stakeholders, neighbors, and community members in order to understand and discuss community needs, issues, and concerns?

a) Charrettes

b) Stakeholder meetings

c) Small task groups

d) Team meetings

**89)** A building envelope allows moisture to accumulate inside its wall assemblies. Beyond damaging the materials, what health-related risk does this create?

a) Hidden mold growth that can harm occupant health and comfort

b) Elevated radon concentrations

c) Increased carbon monoxide production

d) Higher volatile organic compound emissions from insulation

**90)** Which federal agency is responsible for designating flood hazard areas that LEED references when evaluating site sensitivity?

a) U.S. Department of Agriculture

b) U.S. Environmental Protection Agency

c) Federal Emergency Management Agency (FEMA)

d) National Resources Conservation Service

**91)** Which of the following is not a Minimum Program Requirement (MPR)?

a) Every project must be in a permanent location on existing land

b) Every project must use reasonable LEED boundaries

c) Every project must comply with project size requirements

d) Every project must comply with minimum LEED requirements

**92)** Which of the following is the term that combines all the changes, improvements, issued LEED interpretations, and modifications made to a LEED rating system?

a) Credit interpretation rulings (CIR)
b) LEED interpretations
c) Addenda
d) LEED reference guide

**93)** Which of the following is not an outcome of smart growth?

a) Reduce automobile use and promote public transportation
b) Reuse/restore previously developed sites
c) Protect undeveloped land
d) Decreased floor-to-area (FAR) ratios in the urban areas

**94)** A zero-lot-line project has no available on-site area for stormwater management. What does LEED suggest the project team consider?

a) Requesting an exemption from all stormwater requirements
b) Installing a cooling tower to manage excess water
c) Contacting local authorities to discuss off-site rainwater strategies such as community-wide programs
d) Redirecting all stormwater to the municipal sewer system

**95)** Which of the following is named for the sum of all the areas that are affected by the project's activity in the project site?

a) Building footprint
b) Development footprint
c) Carbon footprint
d) LEED project perimeter

**96)** Which of the following statements is false?

a) In LEED, short-term bicycle storage is typically used by visitors for less than two hours and generally does not provide enclosed parking
b) In LEED, long-term bicycle storage is covered storage that protects bicycles from rain and snow for the use of residents and employees
c) LEED encourages projects to provide on-site showers with a changing facility
d) All the newly built projects must provide short term bicycle storage in order to receive a LEED certification

**97)** Which of the following cannot be considered as a low-impact development (LID) strategy?

a) Use of dry ponds (detention ponds)
b) Use of bioswales and vegetated filters
c) Using materials with higher SR and SRI values
d) Bioretention

**98)** A project team is designing a 50,000 square foot office building but determines through analysis that the actual program needs could be met with 38,000 square feet. The team reduces the building size accordingly. Which concept does this decision represent?

a) Dematerialization
b) Rightsizing
c) Source reduction
d) Adaptive reuse

**99)** In which of the following fixtures is water usage measured in gallons per flush (gpf)?

a) Urinals
b) Aerators
c) Shower heads
d) Sink faucets

**100)** LEED requires projects to share whole-project water use data with USGBC. How long must this data-sharing commitment continue?

a) 1 year or until the first recertification
b) 3 years or until the building is renovated
c) 5 years or until the building changes ownership or lessee
d) 10 years regardless of ownership changes

# PRACTICE TEST 2 – ANSWERS

**A score above 80 would indicate very well preparation for the exam.**

| | |
|---|---|
| 1 | D |
| 2 | D |
| 3 | B |
| 4 | C |
| 5 | C |
| 6 | B |
| 7 | D |
| 8 | C |
| 9 | A and B |
| 10 | A |
| 11 | D |
| 12 | B |
| 13 | C |
| 14 | B |
| 15 | B |
| 16 | B |
| 17 | C |
| 18 | B, C and D |
| 19 | D |
| 20 | A |
| 21 | C |
| 22 | C |
| 23 | B |
| 24 | B |
| 25 | A |

| | |
|---|---|
| 26 | A |
| 27 | B |
| 28 | B |
| 29 | A and C |
| 30 | A, B and C |
| 31 | A |
| 32 | D |
| 33 | A |
| 34 | C |
| 35 | C |
| 36 | D |
| 37 | C |
| 38 | A |
| 39 | D |
| 40 | D |
| 41 | B |
| 42 | C |
| 43 | B |
| 44 | B |
| 45 | D |
| 46 | B |
| 47 | C |
| 48 | C |
| 49 | D |
| 50 | C |

| | |
|---|---|
| 51 | B and C |
| 52 | A and D |
| 53 | A, B and D |
| 54 | B |
| 55 | C |
| 56 | D |
| 57 | B |
| 58 | B |
| 59 | C |
| 60 | D |
| 61 | B |
| 62 | A and B |
| 63 | C |
| 64 | D |
| 65 | C and D |
| 66 | D |
| 67 | B |
| 68 | A |
| 69 | C |
| 70 | B |
| 71 | C |
| 72 | B |
| 73 | A, C and E |
| 74 | B |
| 75 | D |

| | |
|---|---|
| 76 | B |
| 77 | D |
| 78 | A, B and E |
| 79 | A and B |
| 80 | D |
| 81 | B |
| 82 | A |
| 83 | D |
| 84 | A |
| 85 | D |
| 86 | B |
| 87 | A and D |
| 88 | B |
| 89 | A |
| 90 | C |
| 91 | D |
| 92 | C |
| 93 | D |
| 94 | C |
| 95 | B |
| 96 | D |
| 97 | C |
| 98 | B |
| 99 | A |
| 100 | C |

# PRACTICE TEST 2 ANSWERS & EXPLANATIONS

**1) Correct answer:** D. Annually.

In LEED v5, outdoor water use is calculated annually, based on the site's plant and irrigation types, using the U.S. EPA's theoretical irrigation requirement (TIR) methodology.

**2) Correct answer:** D. Sorting construction materials into streams for recycling.

Sorting construction materials into streams for recycling is encouraged; however, it is not mandatory for LEED projects.

On the other hand, every LEED project must have a recycling program. This program must cover organics/food waste, mixed paper, corrugated cardboard, glass, plastics, and metals. In addition, all LEED projects must take appropriate measures for the safe collection, storage, and disposal of batteries, mercury-containing lamps, and electronic waste.

**3) Correct answer:** B. Water Efficiency and Energy and Atmosphere.

Water submetering directly supports Water Efficiency by enabling the project to monitor, identify, and reduce water consumption. Optimizing cooling tower operations based on metered data also benefits Energy and Atmosphere because cooling towers are significant energy consumers — reducing their water waste through better blowdown management improves overall system efficiency and reduces the energy required for water treatment and pumping.

**4) Correct answer:** C. Design and construction.

There are two types of review options for LEED BD+C and LEED ID+C projects. The first review option is the **combined review**, in which documentation for all the design and construction prerequisites/credits are submitted for review at the end of the construction phase. At the end of combined review, GBCI will mark the submitted prerequisites/credits as **awarded** or **denied** (this phase is called the **preliminary review phase**). If there is more information needed by GBCI about a prerequisite/credit, GBCI will mark it as clarify, and the project teams will submit the requested clarification (this phase is called the **final review** phase). It is a little riskier compared to the second review option, which is the split review. If some of the expected credits are not awarded, there will not be a make-up, since the project is already completed and no change in design or construction can be made. If a prerequisite is not awarded, then the project will not be able to receive a LEED certification at all.

The second option is the **split review** option, <u>in which the design prerequisites/credits are submitted for review during the design phase, and both the additional design prerequisites/credits (if any) and all the construction prerequisites/credits are submitted at the end of the construction phase</u>. When the design review is complete, GBCI will either mark the design prerequisites/credits either as **anticipated** or **denied.** No prerequisite/credit will be awarded during the design phase, since the design will also need to be implemented on-site during the construction phase. If a design prerequisite/credit is marked as anticipated, it means that the project will earn it at the end of construction phase once that design is implemented on-site. If it's marked as denied, the project teams will need to come up with a design alternative rather than proceeding with that design in order to earn the prerequisite/credit. And at the end of construction review, GBCI will this time mark the submitted prerequisites/credits as **awarded** or **denied**. If there is more information needed by GBCI about a prerequisite/credit, GBCI will mark it as clarify, and the project teams will submit the requested clarification.

**5) Correct answer:** C. Increased need for infrastructure and roads.

Compact development strategies result in reduced need for infrastructure and roads. Suburban sprawl is the top reason for the increased need for infrastructure and roads. A family deciding to move away from New York City center to a suburban region, while employed in the city center can be an example of suburban sprawl. This will result in driving long miles/kilometers, and the increased driving of the family will result in increased greenhouse gas emissions resulting from burning fossil fuels. And after people start moving to suburban areas, the cities will need to provide infrastructure to the suburban areas, which will lead to increased need for infrastructure and roads.

**6) Correct answer:** B. Garden space with different vegetation types that is visible to the building occupants but not physically accessible.

Outdoor spaces should be barrier-free and physically accessible in order to qualify under LEED's Accessible Outdoor Space requirements. A garden that occupants can only view does not qualify, regardless of its vegetation.

**7) Correct answer:** D. Certified.

- LEED Certified: 40-49 points (This is the base certification level)
- LEED Silver: 50-59 points
- LEED Gold: 60-79 points
- LEED Platinum: 80+ points

**8) Correct answer:** C. Improving envelope insulation and HVAC efficiency.

In a hot-arid climate, the dominant energy driver is cooling demand due to intense solar heat gain and large day–night temperature swings. Improving the building envelope through better insulation and high-performance glazing reduces the amount of external heat entering the building, which directly lowers cooling loads. At the same time, upgrading HVAC efficiency ensures that the remaining cooling demand is met with significantly less energy use, making it the most impactful combined strategy.

Increasing ventilation rates is generally less effective in this climate because outdoor air is often hot during the day, which can actually increase cooling loads unless carefully controlled through strategies like night flushing. ENERGY STAR appliances and LED lighting do reduce energy use, but their impact is relatively small compared to envelope and HVAC improvements, which address the building's largest end-use energy demand.

**9) Correct answers:** A. Limiting the number of parking spaces. B. Increasing parking rates.

Limiting the number of parking spaces and increasing parking rates would assist in reducing single-occupancy vehicle use as these measures discourage driving alone and encourage commuters to choose alternative transportation such as carpooling, public transit, biking, or walking.

Locating the project further away from alternative transportation and reducing the number of bicycle parking spaces would increase the single-occupancy vehicle use.

**10) Correct answer:** A. Increase in the carbon dioxide levels (CO2).

Increase in the carbon dioxide levels is the biggest contributor to the global climate change.

**11) Correct answer:** D. Diesel.

Alternative fuels are low-polluting fuels like hydrogen, electricity, propane, compressed natural gas, liquid natural gas, methanol or ethanol.

**12) Correct answer:** B. Green cleaning program.

A **green cleaning program** specifies the green building products to be used, chemicals allowed to be used inside the building, training of the cleaning personnel for the use of chemicals and green cleaning practices, indoor pest control plans, and energy efficient cleaning equipment.

**Green purchasing** covers only the purchase of environmentally friendly products.

Developing a **sustainable material purchasing program** can allow choosing environmentally friendly products when buying office papers, computers, furniture, light bulbs, and more. However does not cover the training of the cleaning personnel for the use of chemicals, implementing green cleaning practices, indoor pest control plans and using energy efficient cleaning equipment.

**Green chemistry** is the design of chemical products and processes that eliminate the use and generation of hazardous substances.

**13) Correct answer:** C. Labor cost.

**Soft cost** covers everything needed for developing a project that does not physically contribute to the building. All the management and supervision costs, design costs, permits and taxes can be seen as the soft costs. While all the other costs that physically contribute to the construction, such as labor costs, the cost of construction materials, and equipment, are **hard costs**.

**14) Correct answer:** B. Positive feedback loop.

"A producing B, which in turn produces more of A" is an example of a **positive feedback loop**.

In **negative feedback loops**, a change brings an additional change in the opposite direction. If a room gets warmer than the set temperature, the thermostat will send a signal to the air conditioning, and the air conditioning will stop blowing warm air.

**Open systems** are systems that constantly consume other items, use them, and produce waste at the end. A whole city may be an example of this, as it imports food, energy, and products; however, while a city uses products, it also produces wastes in turn, often in the form of sewage, pollution, and solid waste, which consequently marks the end of the system. There are no feedback loops in the open systems.

Every system is a composition of its individual parts. Let us take the human body as an example, in which cells form tissues, tissues form organs, organs form the organ systems, and organ systems form humans. A lung is made up of cells; however, having only lung cells together will not do anything. The whole system needs to be there. Certain properties emerge as the result of interaction of individual elements; these are called **emergent properties** of the systems, and only systems can have emergent properties, not the individuals. The breathing function of the lung is the emergent property of a whole successful system.

**15) Correct answer:** B. The project loses the ability to claim on-site renewable energy under LEED because the environmental attributes have been transferred.

RECs represent the environmental and non-power attributes of renewable electricity generation. When a project sells its RECs, the right to claim the environmental benefits of that renewable energy transfers to the buyer.

**16) Correct answer:** B. Basis of design (BOD).

The **basis of design (BOD)** describes the information necessary to accomplish the owner's project requirements, and also includes system requirements, project design criteria, standards, and guidelines.

**17) Correct answer:** C. 30%.

Projects with permanent irrigation must reduce their irrigation water requirement by at least 30% from the baseline, which is calculated for the site's annual theoretical irrigation requirement (TIR) using the U.S. EPA's methodology.

**18) Correct answers:** B. Awarded. C. Denied. D. Clarify.

At the end of construction review phase, GBCI will mark the prerequisites/credits as **awarded** or **denied.** If more information needed about a prerequisite/credit, it will be marked as **clarify**, and the project teams will submit the requested clarification.

**19) Correct answer:** D. Efficiently directing runoff into the municipal storm sewer.

Conveying runoff to the storm sewer is precisely what rainwater management seeks to avoid. In LEED, managing rainwater means retaining it on-site in a manner that replicates natural hydrology — infiltrating it into the soil, returning it to the air through evapotranspiration, or collecting it for reuse — rather than treating it as waste to be piped away.

**20) Correct answer:** A. Black asphalt surface.

A solar reflectance (SR) value will show the solar energy that is reflected by a surface on a scale of 0 to 1. A black surface will have a SR of 0, while a white surface will have a SR of 1. And higher the SR, the lower the heat island effect. Therefore, a black asphalt surface will have the lowest SR compared to the other answer choices.

**21) Correct answer:** C. The team must comply with the interpretation because it was published before their project registration date.

Projects are required to adhere to all LEED interpretations published before their registration date, regardless of whether the team was aware of them at the time. Since this interpretation was published in April 2026 and the project registered in June 2026, the interpretation applies to the project and compliance is mandatory.

**22) Correct answer:** C. Conventional Portland cement concrete.

Conventional Portland cement concrete typically has the highest embodied carbon among common building materials due to the energy-intensive manufacturing process of cement, which involves heating limestone to extremely high temperatures and releases significant $CO_2$ both from fuel combustion and the chemical reaction itself. Timber, especially FSC-certified, stores carbon and has relatively low embodied carbon. Recycled steel requires less energy than virgin steel production. Bamboo is a rapidly renewable material with relatively low embodied carbon. This is why LEED v5 specifically targets cement-based products in its embodied carbon reduction strategies.

**23) Correct answer:** B. The estimated annual miles driven by building occupants commuting to and from the project.

In LEED's Location and Transportation category, vehicle miles traveled (VMT) refers to the estimated annual sum of miles traveled by people driving to and from the project.

**24) Correct answer:** B. The United Nations Sustainable Development Goals (SDGs).

The United Nations' 17 Sustainable Development Goals embed the three pillars of sustainability: economic prosperity, social equity and justice, and environmental stewardship. Green building is integral to achieving virtually all 17 — it reduces carbon emissions, helps communities prepare for climate change, and supports equitable access to affordable energy, safe and healthy workplaces, clean air and water, and the natural world. The Paris Agreement addresses climate commitments specifically, and the Montreal Protocol phases down ozone-depleting substances.

**25) Correct answer:** A. Building orientation.

Building orientation has the biggest effect on a building's energy demand.

**26) Correct answer:** A. Treating and pumping water consumes large amounts of energy, often fossil-fueled.

Conveying water between treatment plants and projects takes large amounts of energy — in fact, water treatment and pumping represent the largest energy use for most U.S. municipalities, typically 30 to 40 percent of a town's energy consumption. When that energy comes from fossil fuels, every gallon saved at the building avoids carbon emissions upstream. This relationship is known as the water–energy nexus, and it is why water efficiency and decarbonization goals reinforce each other.

**27) Correct answer:** B. Performance period.

The performance period is the specific timeframe during which LEED O+M project data is collected to demonstrate prerequisite and credit achievement. This data — including energy, water, waste, and transportation performance — must cover a continuous period that reflects actual building operations.

**28) Correct answer:** B. Local code requirements.

The definition of graywater can change depending on a project's location. Some local codes do not allow the use of graywater as toilet flush water. Because of this, it is advised that the project team should first check with the local code requirements before deciding on using graywater. Local code requirements always take precedent over LEED requirements.

**29) Correct answers:** A. Indigenous plants. C. Native plants.

**Native plants** (or **indigenous plants**) are the type of plants that occur and develop naturally in a specific location.

**Adapted Plants** are types of plants that do not occur naturally in a specific location, however, they can nonetheless adapt easily to the climate of the region.

**Invasive plants** are the types of plants that spread and damage the environment by taking over the adjacent existing native and adapted plants.

**30) Correct answers:** A. Health product declarations (HPD). B. Environmental product declarations (EPD). C. Corporate sustainability reports (CSR).

When choosing environmentally friendly products, a project team should look for health product declarations (HPD), environmental product declarations (EPD), and corporate sustainability reports (CSR).

A product with a **Health Product Declaration (HPD)** will provide disclosure about its material ingredients, list of potential chemicals, related concerns, and additional health information. For manufacturers or raw-material suppliers, **Corporate Sustainability Reports (CSR)** will provide information about the manufacturer or raw-material supplier of a product that has been verified to employ sustainable principles during the creation of their products.

The **Environmental Product Declarations (EPD)**, on the other hand, evaluates the entire lifecycle of a product and assess the cost of the product on the environment. Products that contain an EPD will give information about a product's impact on global warming, ozone depletion, water pollution, greenhouse gas emission, human toxicity, and more.

By considering products with transparent information and declarations, project teams can make better decisions.

**Custodial effectiveness assessments** is about cleanliness of a building, which should also be conducted to obtain the occupants' opinions of the green cleaning program during the building operation phase, to see if they are ever exposed to contaminants. For this question, it does not relate to choosing environmentally friendly products.

Project teams should be careful about **greenwashing** when selecting products. Greenwashing refers to the presentation of a product or a material as being more environmentally friendly than it actually is. Environmentally preferable materials should provide accurate and transparent information and declarations to the buyers.

**31) Correct answer:** A. Preconsumer recycled content.

**Preconsumer recycled content** refers to material that is recycled before being used by a consumer.

**Postconsumer recycled content** is the recycled content of a <u>used</u> material. For example, a recyclable printer paper can be send to recycling after being used and can become a part of a new printer paper. Other types of materials with postconsumer recycled content can be aluminum cans, water bottles, most of the glass, wood and steel products, newspapers and more.

**32) Correct answer:** D. Green-e.

**Green-e** is the leading certification program for the green power generation in the United States, and is not relevant to the VOC levels of the materials.

**33) Correct answer:** A. LEED requires projects not to use inherently non-emitting materials.

Inherently non-emitting materials are those that contain very low or no volatile organic compound (VOC) content by their nature, such as stone, concrete, ceramic tile, glass, brick, untreated solid wood, and metal. Because VOCs are known to cause serious harm to the health of construction workers and building occupants — contributing to respiratory issues, headaches, eye irritation, and long-term health risks — LEED actually encourages projects to use inherently non-emitting materials rather than restricting them.

**34) Correct answer:** C. Filtration media.

Entryway systems are items that are used to capture dirt and particulates at the regularly used exterior entrances, such as grilles, grates, slotted systems and rollout mats.

**35) Correct answer:** C. Environmental justice.

**Environmental justice** is the principle that every person deserves equal protection from environmental and health hazards, equal access to decision-making, and equitable access to a healthy and sustainable environment. **Social determinants of health** describe the conditions in which people are born, live, and work. **Inclusive design** focuses on making buildings accessible to all abilities. **Sustainable development** addresses intergenerational resource needs.

**36) Correct answer:** D. The proposed credit should be applicable to all the LEED rating systems.

An Innovation credit does not need to be applicable to all the LEED rating systems, as all the LEED rating systems have different requirements and are designed to serve particular project types.

**37) Correct answer:** C. Energy recovery ventilation (ERV).

An energy recovery ventilation (ERV) system transfers heat — and in some cases moisture — between outgoing exhaust air and incoming outdoor air. In summer, the ERV pre-cools and dehumidifies incoming hot, humid air using the cooler exhaust air. In winter, it pre-heats incoming cold air using the warmer exhaust. This significantly reduces the heating and cooling energy required to condition high volumes of ventilation air without compromising indoor air quality.

Demand-controlled ventilation reduces airflow when spaces are unoccupied but doesn't help when full ventilation is required. Variable air volume (VAV) systems regulate airflow to zones based on demand but don't recover energy. Duct insulation reduces conduction losses but has minimal impact on conditioning outdoor air loads.

**38) Correct answer:** A. Minimum Energy Efficiency prerequisite.

Prerequisites do not award any points; only credits do. Minimum Energy Efficiency is a prerequisite.

**39) Correct answer:** D. Glare control devices.

Glare control devices are items used to control the glare that daylighting creates; therefore, using them will not increase daylighting. Windows, glass doors, skylights, light tubes, and light shelves can all be used to provide natural daylighting in a building.

**40) Correct answer:** D. Islanding.

**Islanding** is the capability of an on-site energy system to disconnect from the main grid and operate independently during an outage. **Load shedding** is when the utility intentionally reduces grid load to prevent system failure. **Demand response** involves reducing electricity use in response to grid signals. **Grid interactivity** refers to a building's ability to adjust its electricity use based on grid conditions — but it does not specifically describe operating independently from the grid.

**41) Correct answer:** B. Reclaimed water.

**Reclaimed water** is the former blackwater that has been treated and purified for reuse, and can even be used for irrigation.

**Graywater** is the untreated household water that did not come into contact with toilet waste. Used water from bathtubs, showers, bathroom wash basins, and water from clothes washers and laundry tubs can be examples of graywater. Graywater may be used as a flush water in toilets or urinals, however this definition can change depending on the local codes.

**Blackwater** is the term to describe the used water that has come into contact with waste. Thus, the water collected from the urinals and toilets can be classified as blackwater.

**Potable water** is water that is approved for human use, which meets or exceeds U.S. Environmental Protection Agency drinking water quality standards (or a local equivalent outside the U.S.).

**42) Correct answer:** C. LEED rating systems are developed by a profit organization.

LEED rating systems are developed by USGBC, a non-profit organization committed to a sustainable future through cost and energy efficient green buildings.

LEED Green Building Rating System is a **voluntary**, **consensus-based rating system** that **provides third-party verification** that a building or community was designed and built to establish energy savings, water efficiency, location efficiency, improved indoor environmental quality, and stewardship of resources and sensitivity to their impacts. LEED addresses all building types through different rating systems and evaluates a building's performance throughout the building's life-cycle.

**43) Correct answer:** B. A monitoring-based commissioning provider (MBCxP).

An MBCxP uses software to monitor real-time data from building automation systems and meters to detect faulty equipment, identify unusual energy consumption patterns, and address issues as they occur. This is the appropriate professional to systematically investigate why a building's EUI is higher than expected.

**44) Correct answer:** B. Nonpoint source pollution.

Nonpoint source pollution is a type of pollution which its source cannot be identified, and is not relevant with the site assessment.

A site assessment evaluates topography, hydrology, climate, vegetation, soils, human use and human health effects as follows:

- **Topography**: Contour mapping, slope stability risks, unique topographic features.
- **Hydrology**: Flood hazard areas, lakes, delineated wetlands, streams, shorelines, rainwater collection and reuse opportunities.
- **Climate**: Site's solar exposure, heat island effect potential, seasonal sun angles, prevailing winds, monthly precipitation and temperature ranges.
- **Vegetation**: Site's primary vegetation types, significant tree mapping, threatened or endangered species, unique habitat, greenfield area and invasive plant species.
- **Soils**: Natural Resources Conservation Service soils delineation (which is a soil survey showing all the different types of soil on-site), U.S. Department of Agriculture prime farmland, healthy soils, disturbed soils, previous development.
- **Human use**: Views, site's transportation infrastructure, adjacent properties, construction materials with existing recycle or reuse potential.
- **Human health effects**: Proximity of vulnerable populations, adjacent physical activity opportunities, proximity to major sources of air pollution.

**45) Correct answer:** D. LEED BD+C: Core and Shell, because at least 40% of the floor area is incomplete.

LEED BD+C: Core and Shell is required when at least 40% of the project's gross floor area is incomplete at certification — meaning it lacks basic finishes or essential MEP systems. With 42% incomplete, this project exceeds the 40% threshold and cannot use New Construction (which requires at least 60% complete). O+M applies to buildings operational for at least one year. ID+C applies to interior renovations with no exterior scope.

**46) Correct answer:** B. U.S. Department of Energy's Commercial Buildings Energy Consumption Survey (CBECS).

However, **U.S. Department of Energy's Commercial Buildings Energy Consumption Survey (CBECS)** is a database used to estimate a building's total energy cost, and cannot directly reduce the lighting costs.

Innovative approaches can be discovered for more efficient lighting. An example of this could be to use **reflective paving materials**, which will help distribute the light across the site; as a result, the number of fixtures needed to illuminate the area will be reduced, as well as the building operating costs.

Installing **occupancy sensors** will shut off light when an occupant is away from his or her desk, and reduce lighting usage.

Providing **individual lighting controls** with adjustable lighting levels will increase occupant comfort and reduce lighting costs. If a big office space does not contain any individual lighting controls, all the lights should be "on", even if there are few people working in the office.

**47) Correct answer:** C. 30,000 sf.

Floor-to-area ratio (FAR) is calculated by dividing the total square feet of a building to the total square feet of the lot of the building. For this question:

$$3 = \mathbf{A} / 10{,}000$$

$$\mathbf{A} = 30{,}000$$

**48) Correct answer:** C. 60%.

For all LEED BD+C: New Construction and Major Renovations projects, at least **60%** of the project's gross floor area must be complete at the time of LEED certification. To be eligible for the LEED BD+C: Core and Shell rating system, on the other hand, at least 40% of the project's gross floor area must be incomplete at the time of certification.

**49) Correct answer:** D. Searching for Addendas.

Addendas can be searched through USGBC's website, not Arc.

**50) Correct answer:** C. Discovery phase.

The discovery phase is the most important phase of the integrated process and includes predesign activities. During this phase, the project team uses tools such as life-cycle costing, life-cycle assessment, and computer modeling to evaluate strategies, identify risks, assess the triple bottom line, choose the project location, and set green building goals early and cost-effectively. The design and construction phase turns those goals into reality. The occupancy phase monitors long-term performance. Commissioning is a verification process, not a phase of the integrated process.

**51) Correct answers:** B. Heat Island Reduction. C. Rainwater Management.

Using open-grid paving systems contributes to heat island reduction and rainwater management by increasing exposed soil surface, which promotes infiltration and reduces heat absorption.

**52) Correct answers:** A. Green infrastructure (GI). D. Low-impact development (LID).

Green infrastructure (GI) and low-impact development (LID) reduce rainwater runoff by replicating natural hydrology and helping to restore the project site's water balance. Hardscape areas, on the other hand, will increase the rain-water runoff. Efficient storm drains on-site can avoid flooding, however will not replicate natural hydrology, since storm drains transfer the rainwater to the municipal storm water system.

**53) Correct answers:** A. Metals. B. Glass. D. Mixed-paper.

All LEED projects should create a recycling program that should include space for organics/food waste, mixed paper, corrugated cardboard, glass, plastics, and metals. Project teams should also take appropriate measures for the safe collection, storage, and disposal of batteries, mercury-containing lamps, and electronic waste rather than recycling them.

**54) Correct answer:** B. Social Equity and Justice.

The three impact areas of LEED are Decarbonization, Quality of Life, and Ecological Conservation and Restoration. Social Equity and Justice is a component of the triple bottom line, which consists of: Economic Prosperity (Profit), Social Equity and Justice (People), and Environmental Stewardship (Planet).

**55) Correct answer:** C. Green Classroom Professional (GCP) certificate.

The Green Classroom Professional certificate is designed specifically for educators and school facility professionals who want to gain a foundational understanding of green building principles in the context of a school setting. It helps them identify what supports or impedes healthy, resource-efficient, and environmentally sustainable learning spaces.

**56) Correct answer:** D. Does not need to have an accreditation.

A LEED Project Administrator does not need to have an accreditation.

**57) Correct answer:** B. U.S. Environmental Protection Agency.

The WaterSense label is developed by the U.S. Environmental Protection Agency.

**58) Correct answer:** B. LEED AP BD+C.

The LEED AP credit awards 1 point for having a LEED AP with a Building Design and Construction (BD+C) specialty as one of the principal participants on the design team. Legacy LEED APs (LEED APs without a specialty) do not qualify for this credit. The selected LEED AP with specialty must hold an active credential at the time of the project's certification review.

**59) Correct answer:** C. Environmental justice.

Environmental justice holds that every person deserves equal protection from environmental and health hazards, equal access to decision-making, and equitable access to a healthy and sustainable environment. A community near an industrial facility experiencing disproportionate health impacts is a classic environmental justice concern. The project team should consider how their project can avoid adding to the burden and ideally contribute to improving conditions. Material circularity and dematerialization address material use. Compact development addresses land use density.

**60) Correct answer:** D. Follow the local codes instead of LEED requirements.

Project teams should always follow the local codes, as local codes, zoning codes and ordinances take precedent over LEED requirements.

**61) Correct answer:** B. To prevent soil loss, protect waterways from sedimentation, and reduce airborne dust.

An Erosion and Sedimentation Control (ESC) plan is designed to prevent soil loss from the construction site, protect nearby waterways from sediment pollution, and reduce airborne dust generated by construction activities. Common ESC strategies include silt fences, sediment basins, stabilized construction entrances, erosion control blankets, vegetated swales, soil stockpile protection, and dust control measures such as wetting exposed surfaces.

ESC plans manage and treat stormwater runoff to remove sediment, not prevent runoff from leaving the site entirely — that would be a stormwater management or volume control strategy.

**62) Correct answers:** A. Project boundaries. B. Project's size.

There are 3 MPRs applicable to every project:

1. **Every project must be in a permanent location on existing land**
2. **Must use reasonable LEED boundaries**
3. **Must comply with project size requirements**

- LEED BD+C and LEED O+M rating systems: Project must include minimum 1,000 square feet (93 square meters) gross floor area
- LEED ID+C rating systems: Project must include minimum 250 square feet (22 square meters) gross floor area

**63) Correct answer:** C. Thermal comfort.

ASHRAE Standard 55-2023 is used for thermal comfort.

**64) Correct answer:** D. Energy and Resources.

Energy and Resources is not a LEED BD+C credit category. Below is the list of the LEED BD+C credit categories:

- Integrative Process, Planning, and Assessments
- Location and Transportation
- Sustainable Sites
- Water Efficiency
- **Energy and Atmosphere**
- Materials and Resources
- Indoor Environmental Quality
- Project Priorities

**65) Correct answers:** C. LEED ID+C: Commercial Interiors. D. LEED O+M: Existing Buildings.

There isn't any LEED rating system called LEED O+M: Core and Shell, or LEED BD+C: Offices. Tenants of a LEED BD+C: Core and Shell office building can pursue a LEED ID+C: Commercial Interiors certification for the interior fit-out of the office space. And after the occupancy phase, they can pursue the LEED O+M: Existing Buildings certification.

**66) Correct answer:** D. LEED ID+C projects do not require recertification.

LEED ID+C projects do not require recertification. This is different from LEED O+M, where recertification every 3 years is necessary because O+M certifications evaluate the ongoing operational performance of a building.

**67) Correct answer:** B. Brownfield site.

A **brownfield site** is a previously developed site that was contaminated with waste or pollution. **Sensitive lands** are ecologically sensitive areas such as prime farmlands, floodplains, habitat, water bodies or wetland.

**68) Correct answer:** A. U.S. EPA 2022 Construction General Permit (CGP).

LEED requires all projects to create and implement an erosion and sedimentation control (ESC) plan that follows the erosion and sedimentation requirements of the **U.S. EPA 2022 Construction General Permit (CGP)**, the EU Taxonomy: DNSH, Pollution Prevention, Item 4 Noise and Dust; or a local equivalent.

**69) Correct answer:** C. Construction indoor air quality management plan.

SMACNA Guidelines describes the necessary control measures to be taken during construction to protect indoor air quality.

**70) Correct answer:** B. Optimizing the building envelope and passive design strategies first, then sizing mechanical systems based on the reduced loads.

The integrative process prioritizes reducing energy demand through passive strategies — such as building orientation, envelope performance, daylighting, and natural ventilation — before sizing active mechanical systems, which allows for smaller, less expensive, and more efficient HVAC equipment. The other options address energy performance but do not reduce demand at the source, and offsetting or specifying efficient equipment alone misses the larger savings achieved by lowering loads first.

**71) Correct answer:** C. Resilience.

**Resilience** refers to a building's ability to prepare for, withstand, respond to, and recover from disruptive events. Maintaining life-support conditions during a 72-hour power outage through passive design and battery storage demonstrates resilience. **Demand response** involves voluntarily reducing electricity use during grid stress events. **Grid interactivity** refers to adjusting electricity use based on grid conditions while still connected. **Load shifting** moves energy consumption to off-peak hours.

**72) Correct answer:** B. LEED impact areas.

Each credit in LEED makes a contribution to one or more of the three LEED impact areas. Credits that most directly address the prioritized impacts are given the greatest weight—meaning they contain more points. For example, a credit that significantly reduces operational carbon emissions will carry more points than a credit with a narrower environmental benefit.

**73) Correct answers:** A. Ammonia. C. Carbon dioxide. E. Propane.

Natural refrigerants include carbon dioxide ($CO_2$), water ($H_2O$), ammonia ($NH_3$), air, and hydrocarbons such as propane, ethane, and butane. Their ozone depletion potential (ODP) is zero, and their global warming potential (GWP) is zero or very low. However, their suitability and energy performance vary depending on the refrigerant, the system design, and the application. In some cases, a natural refrigerant system may require more energy than a system using a conventional refrigerant, while in others it can perform very efficiently.

**74) Correct answer:** B. Brownfield site.

A **brownfield site** is a previously developed site that was contaminated with waste or pollution. A site that is left from an abandoned building which the contamination is not yet known can also be classified as a brownfield site. **Sensitive lands** are ecologically sensitive areas such as prime farmlands, floodplains, habitat, water bodies or wetland.

**75) Correct answer:** D. Local infrastructure and land use.

The Local Infrastructure and Land Use category evaluates adjacent transportation and pedestrian infrastructure, adjacent diverse uses, relevant local sustainability goals, and applicable accessibility codes. Missing sidewalks and lack of transit access are infrastructure deficiencies that fall directly under this category. Human Use and Health Impacts addresses social services and community safety. Occupant Experience addresses indoor environmental conditions. Demographics evaluates population characteristics like income and education.

**76) Correct answer:** B. LEED Volume Program.

LEED **Volume Program** is for organizations planning to certify a large number of new construction buildings or interior spaces based on a common prototype. It streamlines certification by focusing on similarities in building design and construction practices regardless of location.

LEED **Campus Program** is designed for multiple buildings on a shared site under the control of a single entity. It allows project teams to document shared site-related credits once and apply them across all individual building certifications on the campus. It simplifies the certification process for campus settings such as universities, hospitals, corporate campuses, and industrial parks.

LEED Prototype Program and LEED Uniform Programs do not exist.

**77) Correct answer:** D. Receive points by achieving significant, measurable environmental performance using an innovative strategy which is not addressed in the LEED green building rating system.

When applying for the Innovation credit, a project team can receive points by achieving significant, measurable environmental performance using an innovative strategy that is not addressed in the LEED green building rating system.

Project teams can receive only one point for employing a LEED Accredited Professional (AP) with a specialty appropriate to the project under the LEED AP credit. Employing more than one LEED AP will not award additional points.

Project teams can only achieve the pilot credits that are available on the USGBC's pilot credit library at the time the project decides to register for them.

Project teams can achieve exemplary performance points for qualifying credits, not prerequisites. Prerequisites do not award any points.

**78) Correct answers:** A. Land clearing debris. B. Alternative daily cover. E. Hazardous materials.

Under the LEED requirements, hazardous waste, excavated soil, and land-clearing debris are excluded from LEED waste diversion calculations. Additionally, materials sent for alternative daily cover or incineration/energy recovery are treated as waste. Common construction and demolition waste materials to be diverted include concrete, metals, wood, brick, cardboard, gypsum board, ceilings, furniture, and carpet.

**79) Correct answers:** A. Increased energy usage. B. Increased removal of particles and contaminants from the supply air.

Minimum efficiency reporting value (MERV) rates the air filters according to their success in removing particles from the air. Using air filter with higher MERV values will **increase the energy usage** as more energy will be needed to draw the air from denser filters. Higher MERV will also result in **increased removal of particles and contaminants from the supply air**, and will therefore increase the indoor air quality. However, MERV will not affect thermal quality.

**80) Correct answer:** D. Natural gas.

Renewable energy includes solar, wind, wave, biomass, and geothermal power, plus certain forms of hydropower.

**81) Correct answer:** B. They move existing thermal energy from one place to another rather than generating heat.

A heat pump draws thermal energy out of the air, water, or ground into a refrigerant and transfers that heat to where it is needed, rather than creating heat by burning fuel or running electricity through resistance elements. A ground-source heat pump, for example, moves underground heat into an indoor system in winter — and runs in reverse in summer, drawing heat out of the indoor air. Because moving heat takes far less energy than generating it, heat pumps have displaced direct fossil fuel combustion as the most efficient option for heating and cooling spaces and heating water, which is why they are central to building electrification.

**82) Correct answer:** A. Alternative compliance paths (ACPs).

**Alternative compliance paths (ACPs)** enable international projects to earn the appropriate prerequisites / credits by allowing them to meet international standards or their local standards, instead of U.S. based standards.

**83) Correct answer:** D. HVAC equipment.

Energy use in a building is divided into two main categories: process (unregulated) energy and non-process (regulated) energy. **Process energy** refers to the energy consumed by equipment that is not part of the building's core systems, including computers, office equipment, kitchen stoves and refrigerators, cooking and food preparation equipment, washers and dryers, and elevators and escalators. Most of these items operate by being plugged into outlets, with the exception of elevators and escalators. The consumption of process energy results in building process loads.

**Non-process** or **regulated energy**, by contrast, refers to the energy used to condition spaces and maintain comfort and amenities for building occupants. This includes interior and exterior lighting (such as parking garage, security, landscape, and architectural lighting), HVAC systems (heating, cooling, fans, and pumps), hot water heating, toilet exhaust, and parking garage ventilation. The consumption of non-process energy results in building non-process loads. Since HVAC equipment is part of the building's core conditioning systems, it falls under non-process (regulated) energy rather than process energy.

**84) Correct answer:** A. LEED interpretations are administered by GBCI.

LEED interpretations are administered by USGBC, not GBCI.

Unlike other forms of guidance, LEED interpretations are precedent-setting. This means that once a LEED interpretation is published, it applies to all LEED projects—not just the one that requested it. The clarification effectively becomes part of the rating system itself. LEED interpretations are published online in the Addenda database on USGBC's website for general access and are subject to a consensus-based review by USGBC.

Projects are required to adhere to LEED interpretations posted before their registration date. Adherence to interpretations posted after a project registers is optional but encouraged.

**85) Correct answer:** D. Installing drip-irrigation systems.

Drip-irrigation systems can be used to reduce the <u>outdoor</u> water consumption.

**86) Correct answer:** B. Engaging a commissioning provider to verify system performance during design, construction, and post-occupancy.

Commissioning is a systematic quality assurance process that verifies building systems are designed, installed, tested, and operated to meet the owner's project requirements, with involvement from a commissioning provider throughout design, construction, and post-occupancy phases. The other options provide value but are narrower in scope — a one-time audit captures only a single moment, metering reveals consumption without verifying system function, and warranties cover equipment failure rather than overall performance.

**87) Correct answers:** A. ATMs. D. Vending machines.

LEED does not consider ATMs and vending machines as qualifying use types that count toward a project's proximity to community services and amenities. Below is the list showing the qualifying use types.

1) Food retail: Supermarket, grocery.

2) Community serving retail: Convenience store, farmers market, pharmacy, hardware store, other retail.

3) Services: Bank, family entertainment venue, gym, health club, exercise studio, hair care, laundry, dry cleaner, restaurant, cafe, diner.

4) Civic and community facilities: Adult or senior care (licensed), child care (licensed), community or recreation center, cultural arts facility (museum, performing arts), education facility (e.g., K—12 school, university, adult education center, vocational school, community college), government office that serves public on-site, medical clinic or office that treats patients, place of worship, police or fire station, post office, public library, public park, social services center.

5) Community anchor uses: commercial office (100 or more full-time equivalent jobs), housing (100 or more dwelling units).

**88) Correct answer:** B. Stakeholder meetings.

**Stakeholder meetings** are conducted among the project team, stakeholders, neighbors, and community members in order to understand and discuss community needs, issues, and concerns.

**Charrettes** are intense workshops that are generally held in the beginning of the project and during the project milestones, with the participation of the stakeholders, the project team and experts. During charrettes, participants develop strategies, discuss project goals and create a shared vision.

**Small task groups** are held inside each discipline and together with the involvement of consultants, if needed, to explore particular topics, research, and the refinement of ideas. Then, the results can be collaborated with the whole project team.

**Team meetings** are helpful to discuss strategies among team members from different disciplines. For example, the design of a building envelope will have a great impact on the building's HVAC system. The level of insulation that a facade provides will affect the heating and cooling loads of the building, which will affect the size and type of the HVAC system to be used in the project. Thus, in a team meeting, the facade designer and MEP designer can work collaboratively to discover the best design solutions for both the facade and the HVAC system.

**89) Correct answer:** A. Hidden mold growth that can harm occupant health and comfort.

When moisture penetrates the building envelope, it can lead to hidden mold growth inside assemblies where no one sees it — and mold threatens both the comfort and the health of the people inside. This is why moisture control is a core envelope design concern alongside thermal performance. Radon enters from soil, carbon monoxide comes from combustion, and VOC emissions depend on material chemistry; none of them is caused by damp assemblies.

**90) Correct answer:** C. Federal Emergency Management Agency (FEMA).

FEMA designates flood hazard areas, which LEED references when determining whether a project site is located on sensitive land. The U.S. Department of Agriculture defines prime farmland. The U.S. Environmental Protection Agency (EPA) addresses environmental protection and pollution standards. The Natural Resources Conservation Service designates soil map units.

**91) Correct answer:** D. Every project must comply with minimum LEED requirements.

There are 3 MPRs applicable to every project;

1. Every project must be in a permanent location on existing land
2. Must use reasonable LEED boundaries
3. Must comply with project size requirements

**92) Correct answer:** C. Addenda.

Addenda is the term which combines all the changes, improvements, issued LEED interpretations, and modifications made to a LEED rating system.

**93) Correct answer:** D. Decreased floor-to-area (FAR) ratios in the urban areas.

Smart growth is a planning approach that promotes compact, sustainable development to reduce sprawl, protect natural resources, and create vibrant communities. To achieve these goals, municipalities typically increase — not decrease — floor-to-area ratios (FAR) in urban areas, which encourages denser, mixed-use development on smaller footprints and supports more efficient land use.

The core principles of smart growth include protecting undeveloped land (also known as greenfields), reusing and restoring previously developed sites, reducing automobile use while promoting public transportation, developing efficient rainwater management systems, reducing the heat island effect, minimizing light pollution, and providing stewardship of nature and the surrounding environment.

**94) Correct answer:** C. Contacting local authorities to discuss off-site rainwater strategies such as community-wide programs.

For zero-lot-line projects that cannot manage stormwater on-site due to the building covering the entire lot, LEED suggests contacting local or regional authorities to discuss off-site rainwater strategies, such as participation in community-wide rainwater management programs. LEED does not offer blanket exemptions from stormwater requirements. Cooling towers manage process water, not stormwater. Directing all stormwater to the sewer system contradicts LEED's goal of managing rainwater close to its source.

**95) Correct answer:** B. Development footprint.

The **development footprint** refers to the sum of all areas affected by the project's activity on the project site. **Building footprint** describes the area that the building sits on. **Carbon footprint** refers to the measure of greenhouse gas emissions associated with an activity, such as construction, operation, transportation, or construction materials. At last, **LEED project perimeter** is basically the length of the project boundary.

**96) Correct answer:** D. All the newly built projects must provide short term bicycle storage in order to receive a LEED certification.

LEED does not have a prerequisite that requires projects to provide bicycle storage and on-site showers with a changing facility. Instead, LEED has a Transportation Demand Management credit that requires those provisions. However, since it is a credit, it is optional to pursue.

**97) Correct answer:** C. Using materials with higher SR and SRI values.

Low-impact development (LID) is an approach to mimic natural systems and to manage the stormwater as closest to its source. LID strategies include decreasing impervious surfaces and increasing vegetation on-site, and in addition, LID works on slowing down the flow of rainwater out of site and makes sure the rainwater does not get contaminated.

Using materials with higher SR and SRI values contribute to reducing the heat island effect, and is not a low-impact development strategy.

**98) Correct answer:** B. Rightsizing.

**Rightsizing** means designing a building no larger than its actual needs require. A smaller building demands less energy for heating, cooling, and lighting, and uses fewer construction materials. **Dematerialization** reduces the quantity of materials within a building of a given size. **Source reduction** prevents construction waste through precise material sizing. **Adaptive reuse** designs buildings for future flexibility.

**99) Correct answer:** A. Urinals.

Usage of flush fixtures such as toilets and urinals is measured in gallons per flush (gpf). For flow fixtures, such as sink faucets, showerheads, and aerators, the water usage is measured in gallons per minute (gpm).

**100) Correct answer:** C. 5 years or until the building changes ownership or lessee.

USGBC requires all projects to share whole-project water use data at least once per year, and this commitment must continue for five years or until the building changes ownership or lessee — whichever comes first. This is the same duration required for energy data sharing. The commitment ensures USGBC can track actual building performance over a meaningful period.

# Section 4:

## Practice Test 3

**1)** In a city with high air pollution, using air filters with which of the following minimum efficiency reporting values would be the best choice to enhance indoor air quality?

a) 6

b) 7

c) 8

d) 14

**2)** Which of the following is the primary factor of light trespass onto adjacent sites?

a) Glare

b) Uplight

c) Backlight

d) Facade lighting

**3)** Which of the following publicly available amenities is not classified under the "services" category when evaluating a project's proximity to community services and amenities?

a) Dry cleaner
b) Bank
c) Pharmacy
d) Restaurant

**4)** During the integrative process, the project team decides to orient the building to maximize southern exposure and install light shelves on south-facing windows. Which credit categories does this integrated design decision support? (Choose 2)

a) Location and Transportation
b) Energy and Atmosphere
c) Materials and Resources
d) Indoor Environmental Quality

**5)** LEED refers to which of the following standards for fundamental commissioning?

a) ASHRAE Standard 52.2
b) ASHRAE Standard 55
c) ASHRAE Standard 82.1
d) ANSI/ASHRAE/IES Standard 90.1

**6)** A project team prioritizes reducing total installed materials while maintaining performance. Simultaneously, they select materials with low life-cycle emissions but higher manufacturing energy. What strategy combination is being applied?

a) Electrification + embodied energy reduction
b) Material efficiency + embodied carbon reduction
c) Adaptive reuse + climate risk mitigation
d) Active transportation + emissivity optimization

**7)** A building currently uses a natural gas boiler for space heating, a gas-fired water heater, and gas cooktops in the kitchen. The project team replaces all three with electric heat pumps, electric resistance water heating, and induction cooktops. What has the team eliminated from the building?

a) All emissions from purchased electricity
b) All on-site combustion of fossil fuels
c) All embodied carbon from building materials
d) All process water consumption

**8)** Which of the following standards/guides are project teams advised to review, in order to implement a successful integrative process?

a) ANSI Consensus National Standard Guide 2.0 for Design and Construction of Sustainable Buildings and Communities
b) ASHRAE 241-2023
c) U.S. EPA 2022 Construction General Permit (CGP)
d) Illuminating Engineering Society of North America (IESNA)

**9)** For LEED BD+C projects, how many points are awarded under the Project Priorities credit?

a) 1
b) 5
c) 9
d) 10

**10)** Which of the following strategies will not aid in reducing the heat island effect?

a) Providing undercover/underground parking
b) Installing cool roofs
c) Installing clerestory windows
d) Using open-grid paving

**11)** Using salvaged materials will contribute to reduction in the demand for virgin materials. Which of the following is not a type of salvaged material that can be used in new buildings according to LEED requirements?

a) Bricks
b) Doors
c) Hazardous flooring
d) Cabinets

**12)** Which of the following provides guidance on the types of projects that are eligible for LEED certification, protect the integrity of the LEED program, and reduce the number of issues that arise during the certification process?

a) Minimum Program Requirements (MPRs)
b) Alternative compliance paths
c) Pilot credits
d) Addenda

**13)** Which of the following project costs is not a hard cost?

a) Labor costs
b) Equipment costs
c) Management costs
d) Material costs

**14)** A project team is pursuing LEED v5 certification for a new office building. The project earns 78 points, meets the electrification requirements, achieves enhanced energy efficiency thresholds, procures 100% renewable energy, and achieves a 25% reduction in embodied carbon. What certification level will this project receive?

a) Platinum, because it meets all decarbonization thresholds
b) Gold, because 78 points falls in the Gold range regardless of decarbonization performance
c) Silver, because the project does not meet all prerequisites
d) Platinum, because meeting all decarbonization thresholds overrides the point threshold

**15)** For all LEED BD+C projects, the use of certain refrigerants in HVAC&R systems is prohibited. However, existing small HVAC&R or other equipment—such as standard refrigerators and small water cooler units containing less than __________ of refrigerant—are exempt from LEED requirements.

a) 0.25 pound (123 grams)
b) 0.5 pound (225 grams)
c) 1 pound (450 grams)
d) 2 pounds (900 grams)

**16)** A building stores electricity in batteries from its on-site solar panels during the day and discharges the stored energy during evening peak demand hours when grid electricity is most carbon-intensive. Which two Energy and Atmosphere concepts does this strategy demonstrate?

a) Retrocommissioning and demand-controlled ventilation
b) Energy storage and load shifting
c) Islanding and passive survivability
d) Thermal mass and natural ventilation

**17)** Which of the following best describes the difference between a closed system and an open system?

a) A closed system uses renewable energy while an open system uses fossil fuels
b) A closed system recirculates resources without producing waste, while an open system consumes resources and produces waste
c) A closed system applies to building operations while an open system applies to building construction
d) A closed system has no inputs while an open system has no outputs

**18)** What does LEED require for projects that have not been substantially electrified?

a) Immediate replacement of all fossil fuel equipment
b) Purchase of carbon offsets to compensate for on-site combustion
c) A commitment to electrify within five years of occupancy
d) A decarbonization plan showing cost savings of designing for decarbonization now versus retrofitting later

**19)** Which term describes the practice of meeting current needs without compromising the ability of future generations to meet their own needs?

a) Environmental justice
b) Regenerative design
c) Sustainable development
d) Life-cycle thinking

**20)** Which of the following strategies will not aid in sustainable site management?

a) Developing a sustainable site management plan
b) Using nontoxic snow- and ice-removal products
c) Employing natural ventilation
d) Utilizing Integrated Pest Management (IPM)

**21)** Which of the following is not a part of the commissioning (Cx) plan?

a) Goals and objectives
b) Team member roles/responsibilities
c) Operations and maintenance (O&M) plan
d) Basis of design (BOD) reviews

**22)** Which type of carbon emissions is associated with the daily operation of a building?

a) Embodied carbon
b) Embodied energy
c) Biogenic carbon
d) Operational carbon

**23)** For the projects that install a landscape that does not require any permanent irrigation, the period after the first __________ of the new planting will be considered in the LEED requirements, as the newly planted landscape will need more water to become established in that period.

a) One year
b) Two years
c) Three years
d) Four years

**24)** A building's HVAC filters have a MERV 13 rating. What does this indicate about the filtration system's capability?

a) It can remove gaseous pollutants such as formaldehyde and VOCs from the air
b) It can effectively capture fine particulate matter from the air supply
c) It meets the minimum ventilation rate requirements of ASHRAE 62.1
d) It eliminates the need for outdoor air intake in the ventilation system

**25)** LEED allows projects to earn additional points by addressing environmental concerns that are especially important in their geographic area. What are these location-specific opportunities called?

a) Exemplary performance points
b) Innovation credits
c) Project-type credits
d) Regional priorities

**26)** The __________ is for multiple projects that are located on a single campus owned by the same entity.

a) LEED Volume Program
b) Addenda
c) LEED Campus Program
d) Credit Interpretation Ruling (CIR)

**27)** Which of the following LEED BD+C credit categories awards points for pursuing pilot credits?

a) Regional Priority credit category
b) Project Priorities credit category
c) Energy and Atmosphere credit category
d) Pilot credit category

**28)** To save ventilation energy, a project team programs the HVAC system to exchange air more frequently during the building's normal working hours and less frequently overnight, based on the typical occupancy schedule rather than on sensors. What is this strategy?

a) Natural ventilation
b) Scheduled ventilation rates based on typical occupancy
c) Mixed-mode ventilation
d) Energy-recovery ventilation

**29)** Which strategy represents the highest level of the NIOSH Hierarchy of Controls for protecting indoor air quality?

a) Requiring occupants to wear respiratory protection
b) Installing MERV 13 air filters in the HVAC system
c) Physically removing the source of contamination from the building
d) Posting signage to warn occupants about potential air quality hazards

**30)** How many points are allocated across the core credit categories of a LEED rating system, before the additional Project Priorities points are counted?

a) 100 points
b) 104 points
c) 106 points
d) 110 points

**31)** Which of the following LEED BD+C projects should install permanent water meters that can measure the total water consumption for potable and alternative water sources for the building and associated grounds?

a) LEED BD+C: Commercial
b) LEED BD+C: Residential
c) LEED BD+C: Core and Shell
d) All LEED BD+C projects

**32)** Which of the following is the most effective strategy to reduce waste?

a) Reusing existing buildings
b) Developing new buildings by conducting life-cycle assessment
c) Reusing existing materials in new buildings
d) Developing green buildings

**33)** How many levels of LEED certification are there?

a) 3
b) 4
c) 5
d) 6

**34)** Which strategy best contributes to both heat island reduction and rainwater management?

a) Installing high-albedo roofing
b) Using open-grid pavement with vegetation
c) Installing underground detention tanks
d) Increasing building insulation

**35)** Which of the following strategies is addressed both by the Indoor Environmental Quality credit category and the Energy and Atmosphere credit category?

a) Maximizing building views
b) Using natural ventilation where appropriate
c) Enhancing building acoustics
d) Using low-emitting materials during construction

**36)** An existing building has never been commissioned since it was constructed 15 years ago. The facility manager wants to investigate why the HVAC system is underperforming. Which process should be initiated?

a) Enhanced commissioning
b) Monitoring-based commissioning
c) Retrocommissioning
d) Fundamental commissioning

**37)** __________ refers to the measure of greenhouse gas emissions associated with an activity, such as construction, operation, transportation, or construction materials.

a) Carbon neutrality
b) Development footprint
c) Carbon offsets
d) Carbon footprint

**38)** For fundamental commissioning, LEED requires the qualified commissioning provider (CxP) to have direct commissioning experience on at least ________ projects of equal or larger scope and complexity.

a) One
b) Two
c) Three
d) Five

**39)** Because RECs can be bought and sold separately from the physical electricity, LEED requires that all RECs or EACs associated with the project's renewable energy be ________.

a) Stored for future reporting
b) Sold to the local utility
c) Retired on behalf of the LEED project
d) Transferred to the building owner

**40)** ___________ ensures that the building envelope is tested and verified to achieve the target building performance with respect to energy, water, indoor environmental quality, and durability. This also increases the occupants' comfort levels through infiltration testing, reduced solar heat gain, and glare control.

a) Building envelope commissioning
b) Building exterior commissioning
c) Building envelope checklist
d) Commissioning

**41)** Which of the following is an intentional reduction in the electricity usage in response to a curtailment event or changes in the price of electricity?

a) Retrocommissioning
b) Demand response
c) Green pricing
d) Load shifting

**42)** When using the name of LEED's creator for the first time on any first reference, it should be written as:

a) U.S. Green Building Council
b) USGBC
c) United States Green Building Council
d) U.S.G.B.C.

**43)** __________ are those in which the building footprint extends all the way to one or more property (lot) lines, leaving little or no space for setbacks on those sides, except for limited access areas.

a) Net-zero energy project
b) Regenerative design project
c) Zero-lot-line project
d) Green building project

**44)** Which of the following is a type of building design that creates no waste and also provides more output than consumed input?

a) Regenerative design
b) Sustainable design
c) Integrated design
d) Green design

**45)** A 100,000 square feet of land that has a floor-to-area ratio (FAR) of 2 can allow the construction of a __________ building.

a) 50,000-square-foot
b) 100,000-square-foot
c) 200,000-square-foot
d) 400,000-square-foot

**46)** A building's on-site renewable energy system generates 500 MWh annually. The project team sells 200 MWh worth of RECs to a neighboring property. How much renewable energy can the project claim under LEED?

a) 500 MWh, because the electricity is generated on-site
b) 300 MWh, because 200 MWh of environmental attributes were transferred
c) 0 MWh, because selling any RECs disqualifies the entire system
d) 200 MWh, because that is the amount sold

**47)** A thermostat detects that a room has exceeded its temperature setpoint and signals the air conditioning system to activate. Once the room cools to the desired temperature, the system shuts off. Which systems thinking concept does this illustrate?

a) Positive feedback loop
b) Leverage point
c) Emergent property
d) Negative feedback loop

**48)** A project team identifies that a small change to the building's orientation during early design would significantly reduce both heating loads and electric lighting demand. In systems thinking, what is this early design opportunity called?

a) Emergent property

b) Negative feedback loop

c) Closed system

d) Leverage point

**49)** GBCI has created the __________ credential for professionals who want to demonstrate green building knowledge and expertise in nontechnical fields of practice. It has been created to denote basic knowledge of green design, construction, and operations.

a) LEED Green Associate

b) LEED AP BD+C

c) LEED AP ID+C

d) LEED Fellow

**50)** Which of the following is a voluntary program that sets standards for wood product manufacturers to ensure responsible forest management?

a) Forest Stewardship Council

b) U.S. Housing and Urban Development

c) U.S. Department of Agriculture

d) Land Trust Alliance

**51)** __________ is a process whereby rainwater is collected and filtered to be reused as an alternative to potable water. The filtered rainwater can be used for irrigation, for toilet and urinal flushing, or even for process water in cooling towers.

a) Bioswale

b) Rainwater harvesting

c) Wet pond

d) Dry pond

**52)** For eligible LEED credits, a performance beyond the LEED credit requirements will earn projects __________ points, which are generally achieved by doubling the credit requirements and/or achieving the next incremental percentage threshold.

a) Innovation

b) Exemplary performance

c) Bonus

d) None of the above

**53)** A project team is comparing the costs associated with their LEED project. The architect's design fee is $150,000, the building permit costs $25,000, and the structural steel package costs $500,000. Which of these is classified as a hard cost?

a) The architect's design fee
b) The building permit
c) The structural steel package
d) All three are hard costs

**54)** A facility manager completes a waste stream audit, develops a waste reduction plan, implements changes, and then measures the results. Which operations and maintenance concept does this sequence represent?

a) The integrative process
b) The commissioning cycle
c) A continuous improvement cycle
d) The discovery phase

**55)** Which of the following is not a strategy for designing adaptable, flexible spaces?

a) Designing a flexible floor plan to accommodate future layout changes
b) Using movable wall partitions to allow for floor plan modifications
c) Designing for no interstitial space in order to increase net floor height
d) Using modular systems to accommodate future renovations

**56)** Which of the following organizations develops lighting specifications?

a) Green-e
b) IESNA
c) FSC
d) C2C

**57)** Which of the following cannot be considered as a built environment?

a) Roads
b) Buildings
c) Neighborhoods
d) Wetlands

**58)** Which of the following values are used to calculate the solar reflectance index (SRI) value? (Choose 2)

a) Solar reflectance
b) Emissivity
c) Albedo
d) Heat island effect value

**59)** Locating the project site close to publicly available uses would result in earning credits under which of the following credit categories?

a) Sustainable Sites

b) Location and Transportation

c) Water Efficiency

d) Project Priorities

**60)** A LEED v5 BD+C: Core and Shell project is pursuing Platinum certification. The team knows that New Construction Platinum requires 5 points in Electrification, 8 points in Enhanced Energy Efficiency, 100% renewable energy, and 20% embodied carbon reduction. Can the Core and Shell project assume these same thresholds apply?

a) Yes, because all LEED BD+C adaptations share identical Platinum thresholds

b) No, because Core and Shell projects are exempt from Platinum decarbonization requirements

c) No, because Core and Shell has its own Platinum thresholds in similar decarbonization credits that may differ from New Construction

d) Yes, but only if the project achieves 90 points or more

**61)** Which of the following is the biggest contributor to carbon dioxide emissions?

a) Buildings

b) Transportation

c) Industrial

d) Retail

**62)** Which of the following terms is a measurement unit that describes the building's energy usage relative to its size?

a) Efficiency

b) Energy use intensity

c) Kilowatt-hour

d) Energy factor

**63)** Implementing which of the following strategies is not mandatory for all LEED BD+C projects?

a) Creating an ESC plan according to the erosion and sedimentation requirements of the U.S. EPA 2022 Construction General Permit (CGP)

b) Reducing the irrigation water requirement by at least 30% from the calculated baseline for the site's annual theoretical irrigation requirement (TIR), for projects with permanent irrigation

c) Reducing indoor water usage 30% by utilizing more efficient fixtures and fittings

d) Providing dedicated areas that can be accessible by the waste haulers and building occupants for the collection and storage of recyclable materials for the entire building

**64)** Which of the following types of exterior lighting is not exempt from LEED requirements on uplight and light trespass?

a) Building entrance lighting
b) Directional lightings in transportation
c) Roadway lightings
d) National flag lighting

**65)** Which tool does LEED reference to prioritize adding on-site tree canopy in communities that need it most?

a) Walk Score
b) Global Forest Watch
c) American Forests Tree Equity Score
d) WaterSense Water Budget Tool

**66)** Which of the following strategies will not increase the occupants' comfort levels?

a) Providing thermal comfort controls to the individual occupant spaces
b) Providing lighting controls to the individual occupant spaces
c) Providing quality views
d) Installing high-efficiency toilets

**67)** Which of the following statements is false about a building product with a health product declaration (HPD)?

a) Project teams can see the list of potential chemicals inside the product
b) Project teams can see the material ingredients of the product
c) Project teams can see the health information and related concerns of the product
d) Project teams can see the recyclable materials inside the product

**68)** A project team is planning to install a graywater system in order to use graywater as toilet flush water and earn points in LEED. However, the local code does not allow the use of graywater. What should the project team do?

a) Submit a credit interpretation ruling (CIR) and ask for a clarification
b) Inform GBCI about the local codes and still earn points without installing the graywater system
c) Install the graywater system and inform the local authorities about the reasons for not obeying the local code
d) Follow the local codes instead of LEED requirements

**69)** How many points can a LEED BD+C project receive by employing one LEED AP ID+C and one Legacy LEED AP as principal participants of the project team?

a) 0
b) 1
c) 2
d) 3

**70)** Which of the following statements is false about green building development?

a) Green building projects can qualify for structural incentives such as reduced review and permitting duration

b) If a green building developer is unfamiliar with green building practices, it is better not to design for a green building

c) A green building project can have increased floor-to-area ratio upon certification or proof of green building practices

d) Some municipalities offer reductions or waivers for permit review or other less onerous permitting processes to green buildings

**71)** Which of the following is not a rainwater management strategy?

a) Installing vegetated roofs

b) Using green infrastructure technologies

c) Using open-grid paving

d) Installing cool roofs

**72)** Which of the following strategies cannot aid in reducing the potable water used in cooling towers?

a) Using graywater in cooling towers

b) Increasing the level of treatment to increase the cooling tower cycles using the same condenser water

c) Using reclaimed water in cooling towers

d) Increasing the evaporation rate of the potable water used

**73)** A project team is specifying electric vehicle charging stations. Which two standards does LEED reference for electric vehicle supply equipment requirements?

a) ASHRAE 62.1 and ENERGY STAR

b) National Electric Code (NFPA 70) and ENERGY STAR

c) FEMA flood maps and the IUCN Red List

d) EPA Construction General Permit and WaterSense

**74)** Which of the following strategies is not related with the Sustainable Sites credit category?

a) Avoiding development on sensitive lands

b) Conducting a flush-out

c) Minimizing project disturbances to the neighboring properties and rainwater systems

d) Reducing the pollution created from construction activities by controlling soil erosion and airborne dust

**75)** A 50-story mixed-use development project contains a healthcare facility on the first 10 floors and residential units on the floors above 10. If the gross floor area is the same on each story, which project-type-specific requirements within prerequisites and credits must the project team follow?

a) Only healthcare requirements

b) Only residential requirements

c) The project team can either choose healthcare or residential requirements

d) Both the healthcare and residential requirements must be met

**76)** When addressing the LEED requirements for the heat island effect, which of the following values do project teams need to use for the roofing materials?

a) Albedo value

b) Solar reflectance (SR) value

c) Solar reflectance index (SRI) value

d) Aged solar reflectance (SR) value

**77)** Which of the following strategies will not aid in establishing a good standard of indoor air quality for building occupants following the construction period?

a) Meeting all applicable control measures of the Sheet Metal and Air Conditioning National Contractors Association (SMACNA) IAQ Guidelines during the construction phase

b) Protecting absorptive materials that are installed or stored on-site from moisture damage

c) Changing the filters of the HVAC equipment before the occupancy phase, if the HVAC system is used during construction

d) Implementing a recycling plan for the generated construction waste

**78)** LEED refers to the U.S. Environmental Protection Agency Construction General Permit (CGP) for:

a) Erosion and sedimentation control (ESC) plan

b) Indoor air quality

c) Low-emitting materials

d) Quality views

**79)** Pursuing which of the following may result in earning exemplary performance points if the requirements are met?

a) MPRs

b) Prerequisites

c) Credits

d) None of the above

**80)** Which of the following refers to a system in which all the recyclable materials are mixed in a single collection truck, instead of being sorted into separate commodities such as plastics, glass, and metals?

a) Source separation
b) Commingled recycling
c) Waste-to-energy conversion
d) Source reduction

**81)** LEED requires projects to maintain indoor air quality through several measures during construction. Which of the following is a pathway interruption strategy?

a) Installing low-emitting paints and adhesives
b) Using temporary barriers to isolate high-dust activities from completed spaces
c) Replacing MERV filters before occupancy
d) Conducting occupant comfort surveys after move-in

**82)** Which of the following statements about LEED rating systems is false?

a) The LEED BD+C: Core and Shell Development rating system is eligible for precertification
b) LEED O+M certifications expire every three years and must therefore be recertified
c) LEED ID+C rating systems require in-field verification
d) The LEED BD+C: New Construction and Major Renovations rating system is not eligible for precertification

**83)** Electricity can be generated from ________ resources such as wood waste, landfill gas, agricultural crops or residues, animal waste, and other organic materials.

a) Biomass
b) Diverted materials
c) Low-emitting materials
d) Hide products

**84)** Starting from the predesign and continuing throughout all the design phases, LEED encourages project teams to implement an __________ among building and site systems that will enable project teams to discover unique opportunities to enhance project performance and environmental benefits.

a) Integrative process
b) Triple bottom line process
c) Life-cycle assessment process
d) Commissioning process

**85)** A project team wants to specify bird-friendly glass for the building facade to reduce bird collisions. Which resource does LEED reference for this purpose?

a) NatureServe conservation status database
b) IUCN Red List of Threatened Species
c) American Bird Conservancy Threat Factor Database
d) U.S. Fish and Wildlife Service migratory bird maps

**86)** A project team installs a high-efficiency HVAC system with demand-controlled ventilation that adjusts outdoor air intake based on real-time carbon dioxide levels. Which credit categories does this system support?

a) Energy and Atmosphere and Indoor Environmental Quality
b) Water Efficiency and Sustainable Sites
c) Materials and Resources and Indoor Environmental Quality
d) Location and Transportation and Energy and Atmosphere

**87)** In which of the following integrated process phases should the project teams assess the three pillars of sustainability for the project?

a) Discovery phase
b) Design and construction phase
c) Occupancy, operations, and performance feedback phase
d) Implementation phase

**88)** Which of the following strategies will not aid in reducing single-occupancy vehicle use?

a) Promoting public transportation
b) Limiting the amount of parking spaces in buildings
c) Providing discounted parking rates for electric vehicles
d) Promoting a bicycle-friendly design

**89)** Which of the following strategies cannot be considered a bicycle-friendly design strategy?

a) Providing short-term and long-term bicycle storage spaces for building occupants and visitors
b) Locating the project inside a priority development area
c) Providing on-site showers with a changing facility
d) Locating the project closer to bicycle networks

**90)** Which of the following strategies is addressed both by the Sustainable Sites credit category and the Energy and Atmosphere credit category?

a) Installing solar panels on the roof
b) Using mixed-mode ventilation
c) Reducing parking footprint
d) Utilizing integrated pest management

**91)** Which of the following is the education portal of USGBC?

a) EDUCATION@USGBC
b) Arc
c) USGBC@EDU
d) Greenbuild International Conference & Expo

**92)** Which of the following is the term that is used to define undeveloped land?

a) Greenfield
b) Brownfield
c) Infill site
d) Zero-lot-line land

**93)** A building envelope must manage four different flows to protect the interior. Which component resists the flow of air?

a) Insulation
b) Air-barrier materials
c) Waterproofing
d) Vapor-retarding materials

**94)** A project team is evaluating two sites. Site A has a Walk Score of 88. Site B has a Walk Score of 42. What does this comparison tell the team about these locations?

a) Site A has more parking available than Site B
b) Site A has better access to nearby amenities within walking distance than Site B
c) Site B has better access to nearby amenities within walking distance than Site A
d) Site B is closer to public transit than Site A

**95)** Why does LEED refers to the Sheet Metal and Air Conditioning National Contractors Association?

a) To describe the necessary control measures to be taken during the construction phase to maintain sufficient indoor air quality
b) To reduce the energy usage of the building
c) To reduce the construction and demolition waste disposed of in landfills
d) To encourage the use of products and materials that contain life-cycle information

**96)** Which of the following is not a project type recognized by LEED v5 BD+C for project-type-specific requirements within prerequisites and credits?

a) Data centers
b) Mixed use
c) Schools
d) Warehouses and distribution centers

**97)** A project team harvests rainwater on-site and uses it for both landscape irrigation and toilet flushing. Which credit categories does this strategy address?

a) Sustainable Sites and Water Efficiency
b) Water Efficiency and Energy and Atmosphere
c) Location and Transportation and Sustainable Sites
d) Materials and Resources and Water Efficiency

**98)** Using which of the following types of vegetation will result in using the least amount of fertilizers for landscaping? (Choose 2)

a) Native plants
b) Invasive plants
c) Turfgrass
d) Adapted plants

**99)** MERV stands for:

a) Minimum efficiency requirements value
b) Minimum efficiency replacement value
c) Minimum effectiveness reporting value
d) Minimum efficiency reporting value

**100)** Which of the following statements is false about vegetated roofs?

a) Using vegetated roofs will decrease rainwater runoff
b) Vegetated roofs are much cheaper compared to regular roofs
c) Vegetated roofs are more durable compared to regular roofs
d) Vegetated roofs have great acoustical and insulation qualities

# PRACTICE TEST 3 – ANSWERS

**A score above 80 would indicate very well preparation for the exam.**

| | | | | | | | |
|---|---|---|---|---|---|---|---|
| 1 | D | 26 | C | 51 | B | 76 | C |
| 2 | C | 27 | B | 52 | B | 77 | D |
| 3 | C | 28 | B | 53 | C | 78 | A |
| 4 | B and D | 29 | C | 54 | C | 79 | C |
| 5 | D | 30 | A | 55 | C | 80 | B |
| 6 | B | 31 | D | 56 | B | 81 | B |
| 7 | B | 32 | A | 57 | D | 82 | C |
| 8 | A | 33 | B | 58 | A and B | 83 | A |
| 9 | C | 34 | B | 59 | B | 84 | A |
| 10 | C | 35 | B | 60 | C | 85 | C |
| 11 | C | 36 | C | 61 | A | 86 | A |
| 12 | A | 37 | D | 62 | B | 87 | A |
| 13 | C | 38 | B | 63 | C | 88 | C |
| 14 | B | 39 | C | 64 | A | 89 | B |
| 15 | B | 40 | A | 65 | C | 90 | A |
| 16 | B | 41 | B | 66 | D | 91 | A |
| 17 | B | 42 | A | 67 | D | 92 | A |
| 18 | D | 43 | C | 68 | D | 93 | B |
| 19 | C | 44 | A | 69 | A | 94 | B |
| 20 | C | 45 | C | 70 | B | 95 | A |
| 21 | C | 46 | B | 71 | D | 96 | B |
| 22 | D | 47 | D | 72 | D | 97 | A |
| 23 | B | 48 | D | 73 | B | 98 | A and D |
| 24 | B | 49 | A | 74 | B | 99 | D |
| 25 | D | 50 | A | 75 | B | 100 | B |

# PRACTICE TEST 3 ANSWERS & EXPLANATIONS

**1) Correct answer:** D. 14.

If a project is located inside a city with high air pollution, the filtration level of the supply air needs to be increased, and the air can be further tested to ensure high indoor air quality.

**2) Correct answer:** C. Backlight.

**Backlight** is the primary factor of light trespass onto adjacent sites, **uplight** is the cause of artificial glow, and **glare** is caused by high-angle front lights.

**3) Correct answer:** C. Pharmacy.

Pharmacy is classified under the "community serving retail" category. Below is the list showing the publicly available uses and their categories.

1) Food retail: Supermarket, grocery.

2) Community serving retail: Convenience store, farmers market, pharmacy, hardware store, other retail.

3) Services: Bank, family entertainment venue, gym, health club, exercise studio, hair care, laundry, dry cleaner, restaurant, cafe, diner.

4) Civic and community facilities: Adult or senior care (licensed), child care (licensed), community or recreation center, cultural arts facility (museum, performing arts), education facility (e.g., K—12 school, university, adult education center, vocational school, community college), government office that serves public on-site, medical clinic or office that treats patients, place of worship, police or fire station, post office, public library, public park, social services center.

5) Community anchor uses: commercial office (100 or more full-time equivalent jobs), housing (100 or more dwelling units).

**4) Correct answers:** B. Energy and Atmosphere. D. Indoor Environmental Quality.

Building orientation for maximum southern exposure and the installation of light shelves support Energy and Atmosphere by reducing the need for electric lighting and potentially reducing cooling loads through controlled daylighting. The same strategy supports Indoor Environmental Quality by maximizing daylight penetration into interior spaces, improving occupant comfort and well-being. This is a classic example of how the integrative process creates synergies across credit categories through early design decisions.

**5) Correct answer:** D. ANSI/ASHRAE/IES Standard 90.1.

LEED refers to ANSI/ASHRAE/IES Standard 90.1 for fundamental commissioning.

**6) Correct answer:** B. Material efficiency + embodied carbon reduction.

The team is applying material efficiency and embodied carbon reduction. Material efficiency means using less material to perform the same function — reducing the total amount of material installed while still maintaining the required performance of the project. The second strategy focuses on choosing materials with lower life-cycle greenhouse gas emissions; note that a material can require more manufacturing energy yet still carry lower embodied carbon, which is why emissions, not energy input, are the deciding metric. Electrification addresses building energy systems, not materials, and embodied energy is a different metric from embodied carbon.

**7) Correct answer:** B. All on-site combustion of fossil fuels.

Electrification replaces non-electric energy sources with electricity, eliminating all on-site combustion of fossil fuels. By removing the gas boiler, gas water heater, and gas cooktops, the building no longer burns any fuel on-site, eliminating its direct combustion emissions. Emissions from purchased grid electricity remain until the grid itself runs on clean energy. Embodied carbon relates to materials, not operations.

**8) Correct answer:** A. ANSI Consensus National Standard Guide 2.0 for Design and Construction of Sustainable Buildings and Communities.

LEED encourages project teams to review the ANSI Consensus National Standard Guide 2.0 for Design and Construction of Sustainable Buildings and Communities to implement a successful integrative process.

**9) Correct answer:** C. 9.

The Project Priorities credit awards up to **9 points**.

**10) Correct answer:** C. Installing clerestory windows.

Use of clerestory windows and appropriately sizing and locating all of the windows are important for increasing the amount of daylight introduced into a space. However, this strategy will not reduce the heat island effect.

Below are some strategies to reduce heat island effects:

- Minimize the development footprint (Less development will mean more soil surface.)
- Undercover/underground parking
- Using hardscape materials with high SR values
- Reducing the area of paved surfaces exposed to sunlight
- Installing vegetated roofs (Green roofs are much durable than regular roofs and have great acoustical and insulation qualities.)
- Installing cool roofs (roofs that reflect more sunlight and absorb less heat compared with traditional dark-colored roofs)
- Using roof surfaces with high SRI values (lighter-colored roofs)
- Providing shade from trees, structures covered by solar panels, or architectural devices with high SR values
- Using solar panels on the roof
- Using open-grid paving (In LEED open-grid pavement should be at least 50% permeable.)
- Using cool pavements (pavements that reflect more sunlight and absorb less heat compared with traditional pavements)

**11) Correct answer:** C. Hazardous flooring.

Hazardous materials, such as asbestos, cannot be reused under the LEED requirements, as they cause harm to human health.

Using salvaged materials will contribute to reduction in the demand for virgin materials. Following are some types of salvaged materials that can be used in new buildings:

- Bricks
- Doors
- Windows
- Flooring
- Cabinets, furniture
- Tiles

**12) Correct answer:** A. Minimum Program Requirements (MPRs).

The **minimum program requirements** (**MPRs**) are the minimum standards/characteristics that every LEED project should meet to pursue LEED certification. MPRs state whether a project can pursue LEED certification or not. They provide guidance on the types of projects that are eligible for LEED certification, protect the integrity of the LEED program, and reduce the number of issues that arise during the certification process. They define the types of buildings, spaces, and neighborhoods that a LEED rating system is designed to evaluate.

**Alternative Compliance Path (ACP)** can be an example of a credit adaptation for international projects. ACPs enable international projects to earn the appropriate prerequisites/credits by allowing them to meet international standards or their local standards instead of United States–based standards. Rating systems are also periodically updated to cover more ACPs.

Other than the prerequisites and credits, there are also **pilot credits**, which are credits being tested for the updated version of LEED. According to the feedback received by USGBC from the projects that pursue these credits, these pilot credits can become actual credits of the updated LEED rating system. Pursuing any pilot credit is optional to team members, but if the project chooses to earn a pilot credit, bonus points will be awarded. The points will be awarded under the "Innovation" credit, which is located under the Innovation credit category.

**Addenda** is the term that combines all the changes, improvements, issued LEED interpretations, and modifications made to a LEED rating system. Just like a piece of software updates itself once a week or so to incorporate the latest updates, LEED rating systems are updated with addenda.

**13) Correct answer:** C. Management costs.

The costs during the construction phase can be classified either as soft costs or hard costs. Soft costs cover everything needed for developing a project that does not physically contribute to the building. All the **management** and supervision costs, design costs, permits, and taxes can be seen as soft costs, and all the other costs that physically contribute to the construction, such as labor costs, construction material costs, and equipment costs, are hard costs.

**14) Correct answer:** B. Gold, because 78 points falls in the Gold range regardless of decarbonization performance.

LEED v5 Platinum certification requires BOTH earning 80 or more points AND meeting all specific decarbonization thresholds. Although this project meets all four decarbonization requirements (electrification, enhanced energy efficiency, 100% renewable energy, and more than 20% embodied carbon reduction), it only earned 78 points, which falls in the Gold range (60–79 points). Meeting the decarbonization thresholds alone does not override the point requirement — both conditions must be satisfied simultaneously for Platinum.

**15) Correct answer:** B. 0.5 pound (225 grams).

For existing small HVAC&R or other equipment—such as standard refrigerators and small water cooler units containing less than **0.5 pound (225 grams)** of refrigerant—are exempt from LEED requirements.

**16) Correct answer:** B. Energy storage and load shifting.

The battery system demonstrates energy storage by capturing electricity generated from solar panels for later use. Discharging that stored energy during peak evening hours when grid electricity is dirtiest and most expensive demonstrates load shifting — moving energy consumption from peak to off-peak periods. Retrocommissioning optimizes existing building systems. Demand-controlled ventilation adjusts airflow. Islanding involves disconnecting from the grid during outages. Thermal mass stores heat, not electricity.

**17) Correct answer:** B. A closed system recirculates resources without producing waste, while an open system consumes resources and produces waste.

In systems thinking, a closed system circulates the same medium continuously without producing waste — similar to the cradle-to-cradle concept where materials are endlessly recovered and reused. An open system constantly consumes inputs, uses them, and produces waste — reflecting the traditional cradle-to-grave linear approach.

**18) Correct answer:** D. A decarbonization plan showing cost savings of designing for decarbonization now versus retrofitting later.

LEED requires non-electrified projects to create a decarbonization plan demonstrating the financial case for proactive decarbonization rather than expensive future retrofits. Immediate replacement is not required.

**19) Correct answer:** C. Sustainable development.

Sustainable development is defined as meeting the needs of the present without compromising the ability of future generations to meet their own needs. Environmental justice focuses on equitable protection from environmental hazards. Regenerative design aims to give back more than it takes. Life-cycle thinking considers the full life span of products and buildings.

**20) Correct answer:** C. Employing natural ventilation.

**Employing natural ventilation** is an indoor environmental quality strategy.

Careful plant selection, innovative irrigation systems, and an efficient outdoor lighting design are some strategies that will aid in sustainable site management. During the site management phase, strategies below can be implemented:

- Utilize **Integrated Pest Management (IPM)**, which is known as a sustainable approach that combines knowledge about pests, nature, pest prevention, and control methods that minimize pest infestation and damage while minimizing hazards to the building occupants, the property itself, and the environment. Measures should always be taken to first prevent the attraction of pests. Using native and adapted plantsfor the landscaping can be an effective strategy to combat pests
- **Develop a sustainable site management plan**
- Implement conservation programs with ecologists and nonprofit organizations to protect wildlife species and habitats

- Use nontoxic chemicals for cleaning outside the building
- **Use nontoxic snow- and ice-removal products**
- Carefully use fertilizers and maintenance products in landscaping

**21) Correct answer:** C. Operations and maintenance (O&M) plan.

The O&M Plan is a **separate deliverable** from the Commissioning (Cx) Plan. While both are required project documents, the O&M Plan focuses on how the building will be operated and maintained after occupancy — it is not a component of the Cx Plan itself.

The Cx Plan typically includes:

- Commissioning goals and objectives
- General project information
- Systems to be commissioned
- Cx team member roles and responsibilities
- Owner's Project Requirements (OPR) reviews
- Basis of Design (BOD) reviews
- Functional performance testing procedures
- Verification of system performance
- Procedures for reporting deficiencies and resolving issues

**22) Correct answer:** D. Operational carbon.

Operational carbon refers to emissions produced by running a building on a daily basis — energy used for heating, cooling, lighting, and powering equipment, including both on-site combustion and purchased electricity. Embodied carbon relates to emissions from manufacturing, transporting, and installing building materials, and embodied energy is a measure of energy input rather than emissions. Biogenic carbon is the carbon absorbed and stored by plant-based materials as they grow.

**23) Correct answer:** B. Two years.

For the projects that install a landscape that does not require any permanent irrigation, the period after the **first two years** of the new planting will be considered in the LEED prerequisites and credits, as the newly planted landscape will need more water to become established in the first two years.

**24) Correct answer:** B. It can effectively capture fine particulate matter from the air supply.

MERV (Minimum Efficiency Reporting Value) rates how effectively an air filter removes particles from the air. A MERV 13 filter can capture fine particles including dust, pollen, mold spores, and some bacteria. It does not remove gaseous pollutants like VOCs or formaldehyde — those require activated carbon or other specialized filtration. MERV ratings address filtration effectiveness, not ventilation rates, and outdoor air intake is still required regardless of filter rating.

**25) Correct answer:** D. Regional priorities.

Regional priorities are credits identified by LEED as having particular environmental significance in a project's geographic location. They allow projects to earn bonus points by addressing locally relevant issues. Exemplary performance points are earned by exceeding credit thresholds. Innovation credits reward strategies not covered by existing credits. Project-type credits address needs specific to building types (such as healthcare), not geographic regions.

**26) Correct answer:** C. LEED Campus Program.

**LEED Campus Program** is designed for multiple buildings on a shared site under the control of a single entity. It allows project teams to document shared site-related credits once and apply them across all individual building certifications on the campus. It simplifies the certification process for campus settings such as universities, hospitals, corporate campuses, and industrial parks.

**LEED Volume Program** is for organizations planning to certify a large number of new construction buildings or interior spaces based on a common prototype. It streamlines certification by focusing on similarities in building design and construction practices regardless of location.

**Addenda** is the term that combines all the changes, improvements, issued LEED interpretations, and modifications made to a LEED rating system. Just as a piece of software updates itself once a week or so to incorporate the latest updates, LEED rating systems are updated with addenda.

**27) Correct answer:** B. Project Priorities credit category.

The Project Priority credit under the Project Priorities credit category awards points for pursuing pilot credits.

**28) Correct answer:** B. Scheduled ventilation rates based on typical occupancy.

Optimizing the ventilation rate means exchanging air only as often as necessary for occupants' health, safety, and well-being, and there are two main ways to do it. One is setting different air-exchange rates based on the project's typical occupancy schedule, as this team did. The other is demand-controlled ventilation, which goes further by adjusting airflow in real time using signals from occupancy or air-pollutant sensors — most systems track indoor carbon dioxide levels. Mixed-mode ventilation switches between operable windows and mechanical systems depending on conditions, and energy-recovery ventilation transfers heat between exhaust and incoming air; neither is about when to ventilate.

**29) Correct answer:** C. Physically removing the source of contamination from the building.

The NIOSH Hierarchy of Controls ranks elimination — physically removing the hazard — as the most effective strategy. MERV filters represent engineering controls (isolating people from the hazard). Signage represents administrative controls (changing behavior). Respiratory protection (PPE) is the least effective level. When applied to indoor air quality, eliminating the contaminant source is always the preferred approach over managing exposure after the fact.

**30) Correct answer:** A. 100 points.

Every LEED rating system distributes **100 points** across its core credit categories. The Project Priorities credit category then offers up to 10 additional points, bringing the total possible points on the scorecard to 110.

**31) Correct answer:** D. All LEED BD+C projects.

All LEED BD+C projects should install permanent water meters that can measure the total water consumption for potable and alternative water sources for the building and associated grounds. USGBC also requires all projects to share whole-project water use data with USGBC at least once per year. This commitment must continue for five years, or until the building changes ownership or lessee.

**32) Correct answer:** A. Reusing existing buildings.

The greenest building is the one already built. It can take up to eighty years to make up for the environmental impacts of demolishing an old building and constructing a new one, even if the newly constructed building is environmentally friendly. When reusing existing buildings, it is not necessary to think about waste.

Life-cycle assessment evaluates all the environmental effects of a product during its whole lifetime. A cradle-to-grave, or cradle-to-cradle, approach is used in LCA, and the total energy use and other environmental consequences resulting from the creation of that material are additionally calculated. Hence, conducting an LCA for the whole building is beneficial in showing the trade-offs between different materials and in helping in the selection of materials that would be the best fit for the project and the environment.

**33) Correct answer:** B. 4.

There are **four levels** of LEED certification.

- LEED Certified: 40–49 points (This is the base certification level.)
- LEED Silver: 50–59 points
- LEED Gold: 60–79 points
- LEED Platinum: 80+ points

**34) Correct answer:** B. Using open-grid pavement with vegetation.

Open-grid pavement reduces heat absorption and allows infiltration, addressing both credits. High-albedo roofing only helps heat island. Detention tanks only address stormwater.

**35) Correct answer:** B. Using natural ventilation where appropriate.

Using natural ventilation is addressed both by the Indoor Environmental Quality credit category and the Energy and Atmosphere credit category, as it will increase the indoor air quality of a building in addition to reducing the building's energy usage.

**36) Correct answer:** C. Retrocommissioning.

Retrocommissioning is a systematic process for analyzing and optimizing building system performance in existing buildings, involving a commissioning provider who tests systems and documents measures needed to improve performance. It is the appropriate process for a building that was never commissioned. Enhanced and fundamental commissioning apply during the design and construction of new buildings. Monitoring-based commissioning uses ongoing software-based monitoring and is typically implemented after a building has been commissioned or retrocommissioned.

**37) Correct answer:** D. Carbon footprint.

**Carbon footprint** refers to the measure of greenhouse gas emissions associated with an activity, such as construction, operation, transportation, or construction materials.

**Carbon neutrality** is to emit no more carbon emissions than can realistically be offset.

**Development footprint** is named for the sum of all the areas that are affected by the project's activity in the project site.

A **carbon offset** is a reduction of carbon dioxide made in order to compensate for, or offset, an equivalent carbon dioxide emission made elsewhere. Since energy use of the green building will result in carbon emissions, green buildings can calculate their carbon emissions from their energy use and purchase the same amount of carbon offsets to compensate for their emissions.

**38) Correct answer:** B. Two.

For fundamental commissioning, LEED requires the qualified commissioning provider (CxP) to have direct commissioning experience on at least two projects of equal or larger scope and complexity.

**39) Correct answer:** C. Retired on behalf of the LEED project.

Because RECs can be bought and sold separately from the physical electricity, LEED requires that all RECs or EACs associated with the project's renewable energy be retired on behalf of the LEED project, which prevents double counting.

**40) Correct answer:** A. Building envelope commissioning.

**Building envelope commissioning (BECx)** ensures that the building envelope is tested and verified to achieve the target building performance with respect to energy, water, indoor environmental quality, and durability. This also increases the occupants' comfort levels through infiltration testing, reduced solar heat gain, and glare control.

**41) Correct answer:** B. Demand response.

Think about a power plant that serves the whole city and imagine that the weather gets extremely hot beyond the normal average temperature. As a consequence, everyone simultaneously turns on his or her air conditioner, creating a sudden increase in electricity demand.

If the power plant's energy is not enough to handle that peak demand, then the utility company would think about constructing an additional plant, or it would need to find additional generation sources, including nonrenewable energy sources.

**Demand response** is a technology that aims to overcome these types of conflicts by reducing energy demand, especially during peak times. The utility company sends an alert, which is called a DR event or a curtailment event, to commercial customers who agree to change their usage patterns at peak demands. In turn, the commercial consumers reduce their demand with the alert. Consumers are rewarded for their participation in the demand response , and the construction of additional power plants is avoided. In some cities, utility companies may charge extra during peak times in order to reduce energy demand.

**Retrocommissioning** is a systematic process for analyzing and optimizing building system performance. Retrocommissioning is basically the same process of commissioning applied to existing buildings. The aim of retrocommissioning is to keep a building on track for meeting or exceeding the original operational goals.

If the local utility provider sells green power, the project can directly purchase and use green power. However, if the local utility company does not sell green power, then the project team can pay an extra fee—called **green pricing**—to the local utility company to buy a set quantity of green power instead of buying electricity produced by nonrenewable energy sources.

**Load shifting** is storing the energy generated during off-peak hours to use it during peak-demand hours. Usually, battery systems are used to store the energy during load shifting.

**42) Correct answer:** A. U.S. Green Building Council.

When using the USGBC name on any first reference, it should be written as "U.S. Green Building Council." In subsequent references, the "USGBC" abbreviation can be used. Other naming or abbreviations cannot be used.

**43) Correct answer:** C. Zero-lot-line project.

**Zero-lot-line projects** are those in which the building footprint extends all the way to one or more property (lot) lines, leaving little or no space for setbacks on those sides, except for limited access areas (e.g., entryways, walkways, drive aisles, or service access).

**Net-zero energy project** is a project that only use its own generated renewable energy.

**Regenerative design** is a type of building design that creates no waste and also provides more output than consumed input.

**44) Correct answer:** A. Regenerative design.

**Regenerative design** can be seen to be one step ahead of **sustainable design**, which is a type of building design that creates no waste and provides more output than consumed input. As regenerative design projects do not use any more energy than they produce, they are able to return the excess energy to the utility. In addition to their sustainable features, they can also contribute to the restoration of the lost environment and habitat, thereby taking regenerative design one step ahead of sustainable design.

**45) Correct answer:** C. 200,000-square-foot.

The floor-to-area (FAR) is calculated by dividing the total square feet of a building by the total square feet of the lot of the building. A 100,000 square feet of land that has a FAR of 2 can allow the construction of a **200,000-square-foot building**. If the building has two stories, each story can contain 100,000 square feet of space, which in this case, the building would cover the whole lot since the lot also measures 100,000 square feet. Hence, a higher FAR allows more building space and a denser development.

**46) Correct answer:** B. 300 MWh, because 200 MWh of environmental attributes were transferred.

RECs represent the environmental attributes of renewable electricity. When 200 MWh of RECs are sold, the environmental claim for that portion transfers to the buyer. The project retains the ability to claim only the 300 MWh for which it still holds the RECs. Selling some RECs does not disqualify the entire system.

**47) Correct answer:** D. Negative feedback loop.

A negative feedback loop occurs when a change causes a response in the opposite direction to bring the system back toward balance. The room gets too warm (change), the thermostat activates cooling (response in opposite direction), and the room returns to the setpoint (balance restored). A positive feedback loop amplifies change in the same direction. A leverage point is where a single action creates significant results. Emergent properties arise from the interaction of individual elements within a system.

**48) Correct answer:** D. Leverage point.

A leverage point is a place in a system where a single action can bring about significant, far-reaching results. Changing building orientation during early design is a classic leverage point — one decision simultaneously affects solar heat gain, daylighting, heating loads, and lighting energy across the building's entire life. Emergent properties arise from system interactions. Negative feedback loops restore balance. A closed system recirculates resources without waste.

**49) Correct answer:** A. LEED Green Associate.

GBCI has created the LEED Green Associate credential for professionals who want to demonstrate green building knowledge and expertise in nontechnical fields of practice. It has been created to denote basic knowledge of green design, construction, and operations. LEED Green Associates have a proven, up-to-date understanding of green building principles and practices.

**50) Correct answer:** A. Forest Stewardship Council.

The **Forest Stewardship Council** is a voluntary program that sets standards for wood product manufacturers to ensure responsible forest management in order to prevent deforestation and loss of habitat. The FSC provides a chain of custody (CoC) certification, which is a procedure that tracks a product from the point of harvest/extraction to its distribution.

**U.S. Housing and Urban Development** defines Federal Empowerment Zone, Federal Enterprise Community, and Federal Renewal Community.

**U.S. Department of Agriculture** defines prime farmlands.

**Land Trust Alliance** is an organization that provides accreditation to land trust organizations.

**51) Correct answer:** B. Rainwater harvesting.

**Rainwater harvesting** is a process whereby rainwater is collected and filtered to be reused as an alternative to potable water. The filtered rainwater can be used for irrigation, for toilet and urinal flushing, or even for process water in cooling towers.

**Bioswales** are a stormwater-control feature that uses a combination of engineered basin, soils, and vegetation.

**Wet pond** (retention pond) is a pond designed to hold a specific amount of water indefinitely.

**Dry ponds** (detention ponds) hold the excess rainwater for some time, thereby allowing the rainwater to slowly seep into the ground without contamination. Dry ponds are excavated areas that detain and slow down stormwater but are dry at other times.

**52) Correct answer:** B. Exemplary performance.

For eligible LEED credits, a performance beyond the LEED credit requirements will earn projects **exemplary performance points**, which are generally achieved by doubling the credit requirements and/or achieving the next incremental percentage threshold.

**53) Correct answer:** C. The structural steel package.

Hard costs directly pay for the physical construction of the building, including labor, construction materials, and equipment. The structural steel package is a construction material and therefore a hard cost. The architect's design fee and the building permit are soft costs — expenses needed to plan, design, and manage the project that do not physically become part of the building.

**54) Correct answer:** C. A continuous improvement cycle.

Successful operations and maintenance follows a continuous improvement cycle: assess current conditions through a waste audit, develop strategies in a reduction plan, implement the changes, track the results, and make corrections — then return to assessment and start again. Regular waste tracking keeps the cycle grounded in real data and ensures ongoing performance optimization throughout the building's life. The integrative process focuses on early design collaboration. Commissioning verifies that systems perform as designed. The discovery phase is the first phase of the integrated process, not an ongoing operational cycle.

**55) Correct answer:** C. Designing for no interstitial space in order to increase net floor height.

Designing adaptable, flexible spaces should be well considered in the buildings that may need frequent changes in layouts or floor plans. For example, a flexible floor plan for a hospital building can allow simple changes to be made regarding the sizes of rooms due to the changes in medical equipment that occur alongside the rapid evolution of the technology. Through the use of movable wall partitions or the use of modular systems, the layout of the hospital plans can be changed without going into a complete renovation. With the use of interstitial space, which is an intermediate space between the floors of a building used to run the majority of the building systems, mainly electrical and mechanical, projects can have a much easier time adding or modifying building systems during the building's future adaptation. Therefore, designing with no interstitial space is not a strategy for adaptable spaces. Designing smaller, more flexible buildings with adaptable spaces also makes a structure a better candidate for adaptive reuse — the renovation of an existing building to serve a new use — at the end of its current life.

**56) Correct answer:** B. IESNA.

**Illuminating Engineering Society of North America** (**IESNA**) develops lighting specifications for building design.

**Green-e** is the The leading certification program for the green power generation in the United States.

The **Forest Stewardship Council** (**FSC**) is a voluntary program that sets standards for wood product manufacturers to ensure responsible forest management in order to prevent deforestation and loss of habitat. The FSC provides a chain of custody (CoC) certification, which is a procedure that tracks a product from the point of harvest/extraction to its distribution.

**Cradle-to-Cradle Certification** (**C2C**) assesses the ingredients of a product for environmental and human health hazards and awards a Basic, Bronze, Silver, Gold, or Platinum level Cradle-to-Cradle Certification to the products with preferable life-cycle impacts.

**57) Correct answer:** D. Wetlands.

**Built environment** refers to all the man-made surroundings that are needed for human activity, from roads, to buildings, to neighborhoods.

**58) Correct answers:** A. Solar reflectance. B. Emissivity.

A material's solar reflectance index (SRI) value will indicate a material's ability to stay cool by reflecting solar radiation and emitting thermal radiation. Thus, both the **solar reflectance** and **emissivity** of a material will be combined to rank the material. Emissivity (infrared or thermal emittance) is a measure that shows how much heat or infrared radiation a material can shed back into the atmosphere.

**59) Correct answer:** B. Location and Transportation.

Location and Transportation credit category evaluates the location and the transportation features of the building and their effects on the green building design. Locating the project site close to publicly available uses would result in earning points under the Compact and Connected Development credit.

**60) Correct answer:** C. No, because Core and Shell has its own Platinum thresholds in similar decarbonization credits that may differ from New Construction.

While all LEED v5 BD+C Platinum projects must meet specific decarbonization requirements, the exact thresholds vary by adaptation. Core and Shell, ID+C, and O+M each have their own Platinum thresholds in similar decarbonization-related credits.

**61) Correct answer:** A. Buildings.

Accounting for 38% of carbon dioxide emissions, **buildings** are on top of the list and produce even more carbon dioxide than transportation and industrial emitters. And all the built environment contributes to 67% of all greenhouse gas emissions.

**62) Correct answer:** B. Energy use intensity.

**Energy use intensity (EUI)** is a measurement unit that describes the building's energy usage relative to its size. **Efficiency** is doing the same thing by using less of the same resource. **Kilowatt-hour (kWh)** is a unit of energy equivalent to one kilowatt (1 kW) of power expended for one hour of time. An **energy factor** is a metric used to compare the energy conversion efficiency of residential appliances and equipment.

**63) Correct answer:** C. Reducing indoor water usage 30% by utilizing more efficient fixtures and fittings.

All LEED BD+C projects should reduce their indoor water usage **20%** by utilizing more efficient fixtures and fittings, not 30%.

**64) Correct answer:** A. Building entrance lighting.

In LEED, the uplight and light trespass requirements apply to exterior luminaires within the project boundary. Everyday site lighting such as building entrance lighting must comply.

The following lighting is exempt from these requirements, provided it is controlled separately from the nonexempt lighting:

- Specialized signal, directional, and marker lighting for transportation
- Lighting used solely for facade and landscape lighting in MLO lighting zones 3 and 4, automatically turned off from midnight until 6 a.m.
- Lighting for theatrical purposes for stage, film, and video performances
- Government-mandated roadway lighting
- Lighting for hospital emergency departments, including associated helipads
- Lighting for the national flag in lighting zones 2, 3, or 4

**65) Correct answer:** C. American Forests Tree Equity Score.

LEED references the American Forests Tree Equity Score to identify high-priority and highest-priority areas where on-site tree canopy additions are most needed. Walk Score measures walkability to amenities. Global Forest Watch is available as an alternative for international projects to assess neighborhood tree cover but is not the primary tool for U.S. projects. The WaterSense Water Budget Tool calculates landscape water requirements.

**66) Correct answer:** D. Installing high-efficiency toilets.

High-efficiency toilets (HETs) are WaterSense-labeled toilets, which save 20% of the water typically used in every flush. Even though installing HETs is a great idea to reduce potable water usage, this will not increase the occupants' comfort levels.

Building occupants need to feel comfortable and in control of their environment in order to be healthy, happy, and productive in their buildings. Several interrelated factors influence occupant comfort, including thermal comfort, daylighting, views, and acoustics, all of which shape how a space feels and functions throughout the day. Equally important are the controls that allow occupants to personalize their experience — such as lighting controls and temperature controls — along with ergonomics, which ensures that workstations and furnishings support physical well-being.

However, merely addressing these factors will not be enough. Occupant surveys are also essential to ensure occupant comfort. By receiving occupant feedback, the building management can pinpoint areas that need to be improved.

**67) Correct answer:** D. Project teams can see the recyclable materials inside the product.

**Health product declaration** (**HPD**) is a disclosure that provides a product's material ingredients, list of potential chemicals, related concerns, and additional health information. HPDs will not show the recyclable materials inside product.

**68) Correct answer:** D. Follow the local codes instead of LEED requirements.

The definition of graywater changes from state to state, and some states do not allow the use of graywater as toilet flush water. Because of this, it is advised that the project team first check with the local code requirements before deciding on using graywater.

**69) Correct answer:** A. 0.

The LEED AP credit awards only one point for having a LEED Accredited Professional (AP) with a specialty appropriate to the project as one of the principal participants of the project team. However, for this question, the project cannot receive any points since none of the mentioned LEED APs has the LEED AP BD+C credential. It is also important to note that projects can receive only 1 point under this credit even though they employ more than one LEED AP with a specialty appropriate for the project.

**70) Correct answer:** B. If a green building developer is unfamiliar with green building practices, it is better not to design for a green building.

If a developer is unfamiliar with green building practices, many municipalities provide free planning or certification training and assistance. And in order to promote green developments, many municipalities award structural or financial incentives to developers or homeowners who practice green building techniques.

The structural incentives provide rewards for green building projects by making available additional density bonuses or expedited permitting processes. By reducing the review and permitting durations in exchange for developers' establishment of green building standards, municipalities encourage developers to design green buildings.

Density and height bonuses are another incentive. Many municipalities allow for percentage increases in floor-to-area ratio or other measures of density contingent upon certification or proof of green building practices.

The financial incentives are direct incentives in the form of tax credits or grants to developers that build green buildings. Some municipalities offer reductions or waivers for permit review or other less onerous permitting processes in exchange for developers' use of green building standards.

Revolving loan funds can also be available to those seeking to build or renovate to green building standards.

**71) Correct answer:** D. Installing cool roofs.

Installing cool roofs is a strategy to reduce the heat island effect. Cool roofs cannot absorb rainwater like the vegetated roofs, therefore cool roofs cannot contribute to rainwater management. Cool roofs are roofs that reflect more sunlight and absorb less heat compared with traditional dark-colored roofs.

**72) Correct answer:** D. Increasing the evaporation rate of the potable water used.

If the evaporation rate of the potable water used in cooling towers increases, more potable water will be used.

**73) Correct answer:** B. National Electric Code (NFPA 70) and ENERGY STAR.

For electric vehicle supply equipment (EVSE), LEED references the National Electric Code (NFPA 70) for electrical safety and installation requirements and ENERGY STAR for equipment efficiency standards. ASHRAE 62.1 addresses ventilation. FEMA and IUCN address flood hazards and species conservation. The EPA Construction General Permit addresses erosion control.

**74) Correct answer:** B. Conducting a flush-out.

Conducting a flush-out is a strategy related with the Indoor Environmental Quality credit category. **Flush-out** is the process of supplying ample amounts of fresh air to the building before or during occupancy; this process will take away the contaminated air and establish the desired level of indoor air quality.

**75) Correct answer:** B. Only residential requirements.

Since the gross floor area is the same on each story, the healthcare facility occupies 20% of the total gross floor area (10 out of 50 floors) and the residential units occupy 80% (40 out of 50 floors). In LEED v5, some prerequisites and credits contain requirements that apply to specific project types. If 60% or more of the project's gross floor area consists of a specific project type, the project must follow the requirements for that type within those prerequisites and credits. Since the residential portion exceeds 60%, the project team must follow the residential-specific requirements.

**76) Correct answer:** C. Solar reflectance index (SRI) value.

In order to classify building materials according to their solar emissions and reflectance, the **solar reflectance index (SRI)** is used for the roofing materials, and solar reflectance (SR) values is used for the nonroof materials (such as hardscape) in LEED. In addition to the initial SR and SRI values, aged SR and SRI values will also be needed for LEED credit calculations since the materials' performance will drop as they age.

**77) Correct answer:** D. Implementing a recycling plan for the generated construction waste.

Implementing a recycling plan for the generated construction waste is an effective waste management strategy, however it will not aid in establishing a good standard of indoor air quality for building occupants.

To establish a good standard of indoor air quality for building occupants, project teams can develop and implement an indoor air quality management plan for the construction and preoccupancy phases of the building. The plan must address all of the following:

- During construction, meet or exceed all applicable control measures of the Sheet Metal and Air Conditioning National Contractors Association (SMACNA) IAQ Guidelines.
- Protect absorptive materials that are installed or stored on-site from moisture damage.
- Do not operate permanently installed air-handling equipment during construction unless with filtration media with a minimum efficiency reporting value of 8. The filtration media are installed at each return air grille and return or transfer duct inlet opening, and it is ensured that there is no bypass around the filtration media. Just before the occupancy, all the air filtration media should be replaced with the designed filtration media.
- Prohibit the use of tobacco products inside the building and within 25 feet (8 meters) of the building entrance during construction.

**78) Correct answer:** A. Erosion and sedimentation control (ESC) plan.

With the Construction Activity Pollution Prevention prerequisite, all LEED BD+C projects should establish an **erosion and sedimentation control (ESC) plan** according to the erosion and sedimentation requirements of the U.S. Environmental Protection Agency Construction General Permit (CGP).

**79) Correct answer:** C. Credits.

For eligible LEED credits, a performance beyond the LEED credit requirements will earn projects exemplary performance points, which are generally achieved by doubling the credit requirements and/or achieving the next incremental percentage threshold.

**80) Correct answer:** B. Commingled recycling.

**Commingled recycling** refers to a system in which all the recyclable materials are mixed in a single collection truck, instead of being sorted into separate commodities such as plastics, glass, metals, and more. With commingled recycling, less space is needed to store items, and participation levels may increase since all items go into a single container without sorting.

**Source separation** can be accomplished by sorting the waste materials into multiple recycling streams.

The **waste-to-energy conversion** strategy converts nonrecyclable materials into usable heat, electricity, or fuel through combustion, gasification, pyrolization, anaerobic digestion, and landfill gas recovery (LGR).

**Source reduction** refers to the exact sizing of the materials to be produced through prefabrication, modular construction, or similar methods, in order to prevent waste.

**81) Correct answer:** B. Using temporary barriers to isolate high-dust activities from completed spaces.

Pathway interruption prevents contaminants from migrating to other spaces by isolating high-dust or high-emission activities using temporary barriers, sealed openings, tented work zones, and walk-off mats. Low-emitting materials are a source control strategy. Replacing filters before occupancy is an HVAC protection measure. Occupant surveys are a post-occupancy feedback tool.

**82) Correct answer:** C. LEED ID+C rating systems require in-field verification.

Only LEED Residential projects require in-field verification, not LEED ID+C.

**83) Correct answer:** A. Biomass.

Electricity can be generated from **biomass**- or biofuel-based resources such as wood waste, landfill gas, agricultural crops or residues, animal waste, and other organic materials.

**84) Correct answer:** A. Integrative process.

Starting from the predesign and continuing throughout all the design phases, LEED encourages project teams to implement an **integrative process** among building and site systems that will enable project teams to discover unique opportunities to enhance project performance and environmental benefits.

All green buildings should establish three components: economic prosperity (profit), social equity and justice (people), and environmental stewardship (planet). Those three components are called the **triple bottom line**, also referred to as the three pillars of sustainability. A commitment to the triple bottom line means looking beyond a project's profitability, or considering more than just a project's sustainable features, or social responsibility. When developing a green building, the main goal is to establish all three components at the same time.

**Life-cycle assessment** evaluates all the environmental effects of a product during its whole lifetime. A cradle-to-grave, or cradle-to-cradle, approach is used in LCA, and the total energy use and other environmental consequences resulting from the creation of that material are additionally calculated. Hence, conducting an LCA for the whole building is beneficial in showing the trade-offs between different materials and in helping in the selection of materials that would be the best fit for the project and the environment.

**Commissioning (Cx)** is a systematic investigation by skilled personnel that compares building performance with the project goals, design specifications, and, most importantly, the owner's project requirements (OPR).

**85) Correct answer:** C. American Bird Conservancy Threat Factor Database.

LEED specifically refers to the American Bird Conservancy Threat Factor Database for bird-friendly glass specifications. NatureServe assigns conservation status ranks for species and ecosystems but does not rate glass products. The IUCN Red List identifies endangered species globally but does not address building design. The U.S. Fish and Wildlife Service manages wildlife but is not the specific reference LEED uses for bird-safe glass standards.

**86) Correct answer:** A. Energy and Atmosphere and Indoor Environmental Quality.

A high-efficiency HVAC system directly supports Energy and Atmosphere by reducing energy consumption. Demand-controlled ventilation supports Indoor Environmental Quality by ensuring that adequate fresh air is delivered based on actual occupancy, maintaining good indoor air quality without overventilating. This combination also prevents energy waste from unnecessary ventilation, demonstrating a synergy where one system simultaneously optimizes both energy performance and occupant health.

**87) Correct answer:** A. Discovery phase.

The whole integrative process can be summarized in three main phases:

Discovery phase: This is the most important phase, which also includes predesign to ensure green building goals are set early and cost-effectively. The risks are identified, and the project's three pillars of sustainability — economic prosperity, social equity and justice, and environmental stewardship — are assessed. A system is created that enables the efficient collaboration of every participant in the project. The location of the project is well chosen, and applicable codes and standards are evaluated. The level of LEED certification desired and its requirements are thoroughly considered in this phase. Life-cycle costing, life-cycle assessment, and computer modeling of the project are used to evaluate different strategies and alternatives.

Design and construction phase: The system continues to work, and charrettes, team meetings, small task groups, and stakeholder meetings are continuously conducted. Necessary actions required to protect the construction workers, building occupants, neighborhood, and surrounding environment from construction pollution are taken. The green building goals are turned into reality, with the exception of maintaining them throughout the whole life cycle of the building. All the requirements for LEED certification are documented. The project team conducts both the training and orientation of the building operations personnel.

Occupancy, operations, and performance feedback phase: The performance of all the building systems should be checked throughout the lifetime of the building. The data that the building systems provide are monitored and evaluated. Orientation and training of the building personnel and occupants are conducted repeatedly as tenants and personnel change and new lessons are learned. Building occupants are educated about the green features of the building and sustainable practices. Continuous feedback from all the building users and personnel is received and the system continues for the whole life of the building.

**88) Correct answer:** C. Providing discounted parking rates for electric vehicles.

Providing discounted parking rates for electric vehicles is a strategy to promote green vehicles in order to reduce greenhouse gas emissions resulting from burning fossil fuels. With this strategy, some building occupants may decide to buy a green vehicle, however this wouldn't reduce single-occupancy vehicle use.

**89) Correct answer:** B. Locating the project inside a priority development area.

Locating the project inside one of the high-priority development areas will promote the socially and economically depressed neighborhood.

**90) Correct answer:** A. Installing solar panels on the roof.

Installing solar panels on the roof is addressed both by the Sustainable Sites credit category and the Energy and Atmosphere credit category, as it reduces the heat island effect and generates on-site renewable energy.

Using mixed-mode ventilation is addressed by the Indoor Environmental Quality credit category and the Energy and Atmosphere credit category.

Reducing parking footprint and utilizing integrated pest management are addressed by the Sustainable Sites credit category.

**91) Correct answer:** A. EDUCATION@USGBC.

**EDUCATION@USGBC** is the education portal of USGBC.

**Arc** is the online platform where project teams register their projects for LEED certification, complete prerequisite and credit documentation, and submit applications to GBCI for certification review. It serves as the primary tool for managing the entire LEED certification process from start to finish.

**USGBC@EDU** does not exist.

**Greenbuild International Conference & Expo** is the world's largest expo, launched in 2002, dedicated to green buildings.

**92) Correct answer:** A. Greenfield.

**Greenfield** is the term used to define undeveloped land.

A **brownfield site** is a previously developed site that was contaminated with waste or pollution. A site that is left from an abandoned building in which the contamination is not yet known can also be classified as a brownfield site.

**Infill sites**, or infill developments, are sites that were either previously developed or were already being used for other purposes in urban areas. An example of an infill site could be a project location inside the downtown of a city that was used as a parking lot. Or, it could be the site of a demolished building. Since infill sites have existing infrastructure and also have public transportation options, locating the project at an infill site would not create the negative consequences of suburban sprawl.

Buildings which are built on the entire lot are called **zero-lot-line projects.**

**93) Correct answer:** B. Air-barrier materials.

The envelope works like a fortress against outdoor conditions, and each layer has a job: insulation resists the flow of heat, air-barrier materials resist the flow of air, waterproofing resists the flow of water, and vapor-retarding materials resist the flow of water vapor. These flows must be managed in tandem. Otherwise moisture can get trapped inside the envelope, damaging materials and potentially causing hidden mold growth. A high-performance envelope also increases resilience, keeping the interior comfortable and safe longer during a power outage.

**94) Correct answer:** B. Site A has better access to nearby amenities within walking distance than Site B.

Walk Score measures walkability on a scale of 0 to 100 by evaluating walking routes to nearby amenities. A score of 88 indicates a very walkable location with many amenities accessible on foot, while 42 indicates a car-dependent location. Walk Score does not measure parking availability or transit proximity directly — it specifically evaluates the density and proximity of walkable destinations.

**95) Correct answer:** A. To describe the necessary control measures to be taken during the construction phase to maintain sufficient indoor air quality.

LEED refers to the Sheet Metal and Air Conditioning National Contractors Association (SMACNA) guidelines to describe the necessary control measures to be taken during the construction phase to maintain sufficient indoor air quality.

**96) Correct answer:** B. Mixed use.

LEED v5 BD+C recognizes project types such as schools, data centers, warehouses and distribution centers, healthcare, hospitality, retail, and residential, and certain prerequisites and credits contain requirements written specifically for them. These project type requirements apply when that project type comprises 60% or more of the project's gross floor area. Mixed use is not a project type with its own requirements; it simply describes a building that combines several uses.

**97) Correct answer:** A. Sustainable Sites and Water Efficiency.

Rainwater harvesting supports Sustainable Sites by managing stormwater on-site and reducing runoff that can cause erosion and downstream flooding. Using the harvested rainwater for irrigation and toilet flushing directly supports Water Efficiency by reducing potable water consumption for both outdoor and indoor uses.

**98) Correct answers:** A. Native plants. D. Adapted plants.

**Native plants** (or indigenous plants) are the type of plants that occur and develop naturally in a specific location. Adapted plants are the types of plants that do not occur naturally in a specific location; however, they can nonetheless adapt easily to the climate of the region. In summary, both native and adapted plants can thrive without extensive irrigation, pesticides, and fertilizers.

**Invasive plants** that exist on-site should be cleared from the site, otherwise they will take over the adjacent existing native and adapted plants.Invasive plants are the types of plants that spread and damage the environment.

**99) Correct answer:** D. Minimum efficiency reporting value.

Without quality air filtration media, no building can establish good indoor air quality. The **minimum efficiency reporting value (MERV)** is an important term to know; which rates the air filters according to their success in removing particles from the air.

Air filters with higher MERV values will remove more particles from the air, but more energy will be consumed in passing the air through them. By keeping dust away from HVAC equipment, quality air filters will also increase the performance of the equipment.

**100) Correct answer:** B. Vegetated roofs are much cheaper compared to regular roofs.

Vegetated roofs are more expensive compared to regular roofs, but since they are more durable compared to regular roofs, they can offset this price difference in the long run.

# LEED FUNDAMENTALS

## USGBC

Founded in the early 1990s and based in Washington, DC, the U.S. Green Building Council (USGBC) is a nonprofit organization committed to a sustainable future through cost- and energy-efficient green buildings. USGBC promotes buildings that are environmentally responsible, profitable, and supportive of human health.

USGBC's Mission:

"To transform the way buildings and communities are designed, built, and operated, enabling an environmentally and socially responsible, healthy, and prosperous environment that improves the quality of life."

USGBC's Vision:

"Buildings and communities will regenerate and sustain the health and vitality of all life within a generation."

USGBC accomplishes this mission through the following:

- LEED: The most widely known and used green building program across the world.
- Credentials: The LEED Green Associate credential and the various LEED AP credentials offer professionals a designation to help them stand out in the building industry.
- Greenbuild International Conference & Expo: The world's largest expo, launched in 2002, dedicated to green buildings.
- Education: USGBC provides educational programs about green buildings to further spread environmental principles and design.
- Advocacy: At every level of government, USGBC provides policymakers and community leaders with all the tools, strategies, and resources necessary to inspire action toward a sustainable built environment.
- Chapters & Committees: USGBC maintains a network of regional chapters and community organizations across the country for involvement from the local level to the national level.

# GBCI

**Green Business Certification Inc.** (**GBCI**), formerly known as Green Building Certification Institute (GBCI), changed its name on April 16, 2015; is the sister organization of USGBC that administers LEED building certifications and LEED professional accreditations.

When a project submits all the necessary documents for a LEED application, GBCI is the organization that will review the application and determine if all the requirements set forth in a LEED rating system are fully met or not. And if the requirements are fulfilled, GBCI will grant the LEED certificate.

LEED professional exams are also administered by GBCI, and LEED professional credentials are awarded according to the exam results.

# LEED FOR BUILDING DESIGN AND CONSTRUCTION (LEED BD+C) V5 RATING SYSTEMS

All types of buildings—whether newly constructed or existing buildings undergoing major renovations—will fall under the LEED BD+C category. Major renovations are defined as buildings with extensive interior alteration work in addition to work on the exterior shell and/or primary structural components and/or the core and peripheral mechanical, electrical, plumbing, and service systems and/or site work.

1) LEED BD+C: New Construction and Major Renovations

This is the most widely used rating system under LEED BD+C. It addresses design and construction activities for both new buildings and major renovations of existing buildings. To be eligible, **at least 60% of the project's gross floor area must be complete** by the time of LEED certification. **Complete spaces** are those that have their basic floor, wall, and ceiling finishes installed along with essential mechanical, electrical, and plumbing systems or fixtures necessary to occupy the space for its intended use.

2) LEED BD+C: Core and Shell Development

This is for projects in which the developer controls the design and construction of the building core and the entire mechanical, electrical, plumbing, and fire protection systems, but does not complete the interior tenant fit-out. To be eligible for this rating system, **at least 40% of the project's gross floor area must be incomplete** at the time of certification. **Incomplete spaces** are those that do not have their basic floor, wall, and/or ceiling finishes installed or essential mechanical, electrical, or plumbing systems or fixtures necessary to occupy the space for its intended use.

# LEED FOR INTERIOR DESIGN AND CONSTRUCTION (LEED ID+C) V5 RATING SYSTEM

LEED ID+C rating system applies to interior renovation projects of commercial spaces that have **no scope for the exterior shell of the building or site work**. The team installs movable furnishings, fixtures, and equipment (FF&E) to support the intended regular operations of the space.

1) LEED ID+C: Commercial Interiors

This is for interior spaces undergoing a complete fit-out. If a company rents an office in a building and decides to design and construct its interior office space, the LEED ID+C: Commercial Interiors rating system can be used to certify that space.

# LEED BUILDING OPERATIONS AND MAINTENANCE (LEED O+M) V5 RATING SYSTEM

The LEED O+M rating system applies to existing buildings that are focused on operational improvement. This rating system covers whole buildings that have been **fully operational and occupied for at least one year**. Rather than evaluating design and construction, LEED O+M certifies the ongoing operations and maintenance of the building through performance-driven strategies and outcomes, allowing buildings in use to achieve greater efficiency.

A LEED ID+C: Commercial Interiors certified office located in a LEED BD+C: Core and Shell certified building can also receive a LEED O+M: Existing Buildings certification.

LEED O+M certifications expire every three years, and in order to maintain a LEED O+M certification, a project needs to be **recertified every three years**. Certifications under the other rating systems do not expire and do not require recertification.

1) LEED O+M: Existing Buildings

This is the primary rating system under LEED O+M and applies to existing buildings of all use types that are focused on improving operational performance.

# LEED FOR CITIES V5 RATING SYSTEM

LEED for Cities is a rating system designed to help cities and communities measure, manage, and improve their sustainability performance at a broader scale. Rather than certifying a single structure, LEED for Cities evaluates the environmental, social, and economic performance of an entire city or community.

# MINIMUM PROGRAM REQUIREMENTS (MPR)

The **minimum program requirements** are the minimum standards/characteristics that every LEED project should meet to pursue LEED certification. MPRs state whether a project can pursue LEED certification or not. There are three MPRs:

**1) Every project must be in a permanent location on existing land**

A project that is designed to move at any point in its lifetime is not eligible for LEED certification. Prefabricated or modular structures can be LEED-certified as long as they are installed permanently.

**2) Must use reasonable LEED boundaries**

The LEED project boundary must include all contiguous land that is associated with the project and that supports its typical operations. These will include the landscaping, septic or stormwater treatment equipment, parking, sidewalks, and even more. Any project space cannot be shown to be excluded in order to give the project an advantage in complying with credit requirements.

**3) Must comply with project size requirements**

All LEED projects must meet the following size requirements according to their rating system:

- LEED BD+C and LEED O+M rating systems: The project must include a **minimum of 1,000 square feet (93 square meters) of gross floor area.**
- LEED ID+C rating systems: The project must include a **minimum of 250 square feet (22 square meters) of gross floor area.**

# LEED IMPACT AREAS AND CREDIT POINT ALLOCATION PROCESS

LEED v5 is structured around **three impact areas** which are the overarching objectives that every LEED project aims to accomplish, and every prerequisite and credit in the rating system is connected to one or more of them. The three impact areas are:

**1) Decarbonization (50% of available points)**: This is the dominant impact area in LEED v5, reflecting the urgency of the climate crisis and the building industry's significant contribution to greenhouse gas emissions. Decarbonization targets all major sources of carbon emissions from buildings: operational carbon from energy use, embodied carbon from materials and construction, refrigerant emissions from mechanical systems, and transportation emissions from occupant commuting.

**2) Quality of Life (25% of available points)**: This impact area recognizes that buildings are fundamentally about people. It addresses the health, well-being, resilience, and equity of building occupants and surrounding communities. LEED v5 requires project teams to understand who uses the building, assess the needs of the local community, evaluate natural hazards that could affect the site, and consider the workers and supply chains behind building materials.

**3) Ecological Conservation and Restoration (25% of available points):** This impact area focuses on limiting the environmental degradation caused by building construction and operations and promoting the rehabilitation and restoration of ecosystems. LEED v5 takes a comprehensive approach to ecological health by incorporating strategies that conserve natural habitats, minimize impacts on land and wildlife, restore ecosystems to enhance resilience and biodiversity, and manage natural resources responsibly.

Each credit in LEED makes a contribution to one or more of these impact areas. Credits that most directly address the prioritized impacts are given the greatest weight—meaning they contain more points.

# LEED CERTIFICATION PROCESS FOR LEED BD+C AND LEED ID+C PROJECTS

## 1. PROJECT REGISTRATION

The first step for LEED-certifying a building is to register the project through Arc, which means completing the registration form and paying the **registration fee** that varies by membership status. In this phase the project team should select the appropriate LEED rating system.

The certification platform defines two key roles: the owner and the LEED Project Administrator. The **owner** of the project refers to the individual(s) or entity(ies) that has the legal authority to fulfill the obligations defined in the LEED Certification Agreement. A **LEED Project Administrator** is the primary project contact for GBCI. The LEED Project Administrator is the team member who acts as a project manager, overseeing the LEED project and organizing team members for certain tasks, credits, or prerequisites.

## 2. APPLICATION

After the project registration phase, the project team will identify LEED credits to pursue and assign responsibility for the various credits to project team members. The team will then collect information, perform calculations, prepare documentation, fill out the credit forms demonstrating the achievement of the prerequisites and credits, and upload all the documentation to Arc. Once everything is complete, the LEED Project Administrator will verify the submission and the project team will pay the review (certification) fees to submit the application.

## 3. REVIEW

There are two types of review options for LEED BD+C and LEED ID+C projects: combined review and split review.

<u>**Combined review**</u>:

At the end of the construction phase:

1. Both design and construction submittals are sent by the project team.
1. There's a **preliminary review** by GBCI.
2. If additional information is needed to evaluate the application, GBCI will ask for clarifications and the project team will send the clarifications.
3. GBCI will proceed with the **final review** to evaluate the clarifications and to report the <u>awarded</u>/<u>denied</u> prerequisites and credits to the project team.
4. The project team will either accept or appeal the result.

<u>**Split review**</u>:

During the design phase:

1. Design submittals are sent by the project team.
2. **Preliminary design review** is conducted by GBCI.
3. If additional information is needed to evaluate the application, GBCI will ask for clarification, and the project team will send the clarification.
4. GBCI will proceed with the **final design review** to evaluate the clarifications and report the <u>anticipated</u>/<u>denied</u> prerequisites/credits to the project team.
5. The project team will either accept or appeal the result.

At the end of the construction phase the following occurs:

6. Both construction and extra design submittals (if applicable) are sent by the project team.
7. A **preliminary construction review** is conducted by GBCI.
8. If additional information is needed to evaluate the application, GBCI will ask for clarification, and the project teams will send the clarification.
9. GBCI will proceed with the **final construction review** to evaluate the clarifications and report the <u>awarded</u>/<u>denied</u> prerequisites/credits to the project team.
10. The project team will either accept or appeal the result.

## 4. CERTIFICATION OR DENIAL

This is the last step of the LEED certification. GBCI issues the final certification report, which will show the level of LEED certification awarded unless the certification is denied. The project teams can again appeal the results if needed, and if no appeals are made, then the project will be deemed "**closed out,**" which means the project team will no longer be able to appeal the certification level or review decisions for specific credits or prerequisites.

The total number of earned points will determine the level of LEED certification, assuming all the prerequisites and MPRs are met.

- LEED Certified™: 40–49 points
- LEED Silver: 50–59 points
- LEED Gold: 60–79 points
- LEED Platinum: 80–80+ points

**This point system applies to all the rating systems in LEED.**

LEED v5 Platinum projects must also meet specific decarbonization requirements in designated credits. For LEED BD+C: New Construction projects, for example, Platinum certification requires meeting minimum thresholds in four specific credits:

- Electrification Credit — Five points are required.
- Enhanced Energy Efficiency Credit — Eight points are required.
- Renewable Energy Credit — 100% of site energy use from any combination of Tier 1, Tier 2, and Tier 3 renewable energy.
- Reduce Embodied Carbon Credit — 20% reduction in embodied carbon.

Core and Shell, LEED ID+C, and LEED O+M projects have their own Platinum thresholds in similar decarbonization-related credits.

# LEED CERTIFICATION PROCESS FOR LEED O+M PROJECTS

A LEED O+M certification is not granted for good, unlike the case with the other rating systems. A LEED O+M project needs to get recertification **every three years**. And there isn't a split review option available. The LEED O+M project will only have an initial review, which at the end may result in the grant of the LEED O+M certificate. The rest of the certification process is the same with the LEED BD+C and LEED ID+C projects.

# SUMMARY OF STRATEGIES TO ADDRESS INTEGRATIVE PROCESS, PLANNING, AND ASSESSMENTS

## INTEGRATED BUILDING PROCESS

The integrated process emphasizes connection and communication among all professionals and stakeholders in the project. To see the big picture, everyone involved should act as a single team. Stakeholders, architects, structural engineers, electrical engineers, mechanical engineers, facility managers, and others should work collaboratively starting from the predesign phase. Each building system should be addressed from predesign through rigorous questioning and a deliberate challenge of typical project assumptions.

During predesign or early design, the team conducts a design charrette with the owner (or owner's representative) and participants representing at least four key perspectives. Charrettes are intensive, multidisciplinary workshops held at the beginning of a project and at key milestones, allowing brainstorming and collaboration among different disciplines and experiences.

LEED requires projects to define a set of specific, measurable goals that address the LEED v5 impact areas. These goals should be incorporated into the owner's project requirements and tracked through design via the basis of design and other project documents. The **owner's project requirements** (**OPR**) document the owner's vision and performance requirements, including the LEED objectives. The **basis of design** (**BOD**), developed by the architect/engineer, describes the information necessary to accomplish the OPR, including system requirements, design criteria, standards, and guidelines.

To become familiar with the integrative process, project teams are encouraged to review the ANSI Consensus National Standard Guide 2.0 for the Design and Construction of Sustainable Buildings and Communities.

# PHASES OF THE INTEGRATED PROCESS

The whole integrated process can be summarized by three main phases:

- **Discovery phase**: This is the most important phase, which also includes predesign to ensure green building goals are set early and cost-effectively. The risks are identified, and the triple bottom line of the project is assessed. A system is created that enables the efficient collaboration of every participant in the project. The location of the project is well chosen, and applicable codes and standards are evaluated. The level of LEED certification desired and its requirements are thoroughly considered in this phase. Life-cycle costing, life-cycle assessment, and computer modeling of the project are used to evaluate different strategies and alternatives.
- **Design and construction phase**: The system continues to work, and charrettes, team meetings, small task groups, and stakeholder meetings are continuously conducted. Necessary actions required to protect the construction workers, building occupants, neighborhood, and surrounding environment from construction pollution are taken. The green building goals are turned into reality, with the exception of maintaining them throughout the whole life cycle of the building. All the requirements for LEED certification are documented. The project team conducts both the training and orientation of the building operations personnel.
- **Occupancy, operations, and performance feedback phase**: The performance of all the building systems should be checked throughout the lifetime of the building. The data that the building systems provide are monitored and evaluated. Orientation and training of the building personnel and occupants are conducted repeatedly as tenants and personnel change and new lessons are learned. Building occupants are educated about the green features of the building and sustainable practices. Continuous feedback from all the building users and personnel is received and the system continues for the whole life of the building.

# STRATEGIES TO ADDRESS INTEGRATIVE PROCESS, PLANNING, AND ASSESSMENTS CREDIT CATEGORY

- Implement an integrative design process that brings together diverse stakeholders and specialists from the earliest stages of the project.
- Collaboratively set specific, measurable project goals that align with the LEED v5 impact areas of decarbonization, quality of life, and ecological conservation and restoration, and incorporate them into the owner's project requirements.
- Conduct a carbon assessment early in the design process to understand the project's total carbon footprint.
- Analyze the embodied carbon of building materials and construction activities.
- Evaluate long-term carbon emissions associated with the building's energy consumption.
- Estimate the emissions generated by occupants and visitors commuting to and from the project and explore strategies to reduce transportation-related carbon.
- Conduct a climate resilience assessment to identify natural hazards and climate-related threats relevant to the project's location.
- Perform a climate risk assessment to estimate the potential consequences of identified hazards.
- Develop resilience strategies that allow the building to prepare for, withstand, respond to, and recover from disruptive events.
- Engage community stakeholders early in the planning process to ensure the project reflects local needs and priorities.
- Conduct a human impact assessment to evaluate how the project affects surrounding communities and populations.
- Study the demographics of the surrounding area and identify any vulnerable populations, including fenceline communities, that may be affected by the project.
- Evaluate race and ethnicity, gender, age, income, employment rate, education levels, and household types in the project's surrounding area.
- Review local infrastructure, land-use policies, and zoning conditions that may influence the project's social and environmental impact on the community.
- Evaluate adjacent transportation and pedestrian infrastructure, adjacent diverse uses, relevant local sustainability goals, and applicable accessibility codes.
- Evaluate the availability of natural daylight, quality of views, and the opportunity to provide operable windows.
- Evaluate air and water quality in the project's surrounding area.
- Use the findings of the human impact assessment to inform the project's planning, design, and operations and maintenance.
- Address social determinants of health by designing spaces that improve access to clean air, safe environments, green spaces, and community resources.
- Ensure appropriate workforce protections are in place so that construction crews are not exposed to unsafe working conditions.
- Select building materials from manufacturers that demonstrate fair labor practices and seek certifications that verify worker health and safety protections throughout the supply chain.
- Consider the environmental effects of ongoing purchases related to building operations and maintenance, such as cleaning products, replacement materials, and landscaping supplies.

# SUMMARY OF STRATEGIES TO ADDRESS LOCATION AND TRANSPORTATION

## LOCATION

The location of a green building project should first promote **smart growth,** which refers to the approach that protects undeveloped lands and contributes to developments in projects near jobs, schools, shops, and other destination points with diverse uses. An example may be a residential project that is located very close to downtown, which also contains several public transportation options.

However, a residential project in a suburban area that requires driving many miles to the downtown area for people to commute would not contribute to smart growth, and that project would hence support suburban sprawl.

Principles of smart growth are:

- Protect undeveloped land (undeveloped land is also called **greenfield**)
- Reuse/restore previously developed sites
- Reduce automobile use and promote public transportation
- Develop efficient rainwater management
- Reduce the heat island effect (will be discussed in the next chapter)
- Reduce lighting pollution
- Provide stewardship of nature and the site's surroundings

According to the U.S. Environmental Protection Agency (EPA), in order to promote smart growth and discourage suburban sprawl, the following strategies should be applied:

- Mix land uses (residential, commercial, retail, etc. all on the same land)
- Take advantage of compact building design
- Create a range of housing opportunities and choices
- Create walkable neighborhoods
- Foster distinctive, attractive communities with a strong sense of place
- Preserve open space, farmland, natural beauty, and critical environmental areas
- Strengthen and direct development toward existing communities
- Provide a variety of transportation choices
- Make development decisions predictable, fair, and cost effective

- Encourage community and stakeholder collaboration in development decisions

## PROTECTING THE HABITAT BY CHOOSING THE RIGHT LOCATION

To protect the habitat, green buildings should not be developed inside **sensitive lands**; instead, **infill sites** should be preferred. Infill sites, or infill developments, are sites that were either previously developed or were already being used for other purposes in urban areas.

Below are the types of sensitive lands that should be avoided for development:

- Prime farmland
- Sites close to wetlands and water bodies (lakes, rivers, etc.)
- Public parkland
- Areas below floodplain
- Areas that are a part of the habitat or endangered species

Locating the project on brownfield sites and remediating them before the start of construction can be another option, which can also gain extra points in the LEED certification.

**Brownfield sites** are previously developed sites that were contaminated with waste or pollution. A site that has an abandoned building with unknown contamination can also be classified as a brownfield site. Once the contamination is cleaned up, these sites can be an excellent option for project development, and the federal, state, or municipal government can offer tax incentives on the property to support development.

## TRANSPORTATION

To minimize the environmental impact of transportation and promote sustainable commuting, projects should:

- Complete a transportation demand assessment
- Reduce the consequences of transportation by ensuring access to alternative modes of transportation to reduce single-occupancy vehicle use
- Encourage active transportation, including walking, biking, and wheeling
- Support electric vehicle adoption through the installation of electric vehicle supply equipment (EVSE)
- Take advantage of high-density or infill locations to give occupants walking access to diverse uses and public transit
- Promote local connectivity, EV use, carpooling incentives, and walkable diverse uses when public transit is unavailable
- Utilize compressed workweeks or telecommuting to reduce commuting-related fossil fuel use
- Reward carpooling and increase parking rates to discourage single-occupancy vehicle use

- Limit the number of parking spaces and charge fees at or above the local market rate, as addressed under the Transportation Demand Management credit
- Provide secure bicycle storage, shower rooms, and bike repair stations to support bicycle commuting and physical activity

# NEIGHBORHOOD PATTERN AND DESIGN

Following are the features of a healthy neighborhood (which are also part of **compact development strategies**):

- Wide sidewalks, benches, and bicycle networks
- Business centers, retail services, educational facilities, and other diverse uses, appropriately located to minimize the length of travel
- Easily accessible public transportation options
- Street layouts that allow for easy connectivity
- Community gardens, farmers markets, and agricultural programs to support access to sustainable food

# EQUITABLE DEVELOPMENT AND COMMUNITY IMPACT

Projects should support inclusion, community well-being, and equitable economic opportunity for both existing residents and the broader community:

- Be sensitive to existing residents and avoid making neighborhoods unaffordable or displacing community members
- Set aside affordable units for people earning below the area median income
- Include spaces and services that respond to local needs through early and ongoing community engagement
- Engage with community members, local nonprofits, and public agencies throughout planning and design
- Leverage schools as community resources by opening underused spaces (gyms, auditoriums) for public use and providing space for health clinics or social support services
- Create equitable economic opportunities through construction jobs that offer stable employment and higher wages without requiring a college degree
- Hire workers from the surrounding area and offer apprenticeships, skills training, and pathways for long-term economic advancement
- Extend support beyond the jobsite through programs such as financial literacy training and entrepreneurship education

# SUMMARY OF STRATEGIES TO ADDRESS SUSTAINABLE SITES

## SITE ASSESSMENT

As an important aspect of the integrated process of creating a sustainable site, a site assessment first needs to be conducted in order to have the relevant information that will be necessary for the site design.

The project team can conduct an assessment evaluating the following site features:

- Topography: Contour mapping, slope stability risks, unique topographic features
- Hydrology: Floor hazard areas, delineated wetlands, lakes, streams, shorelines, rainwater collection, and reuse
- Climate: Solar exposure, seasonal sun angles, winds, precipitation values
- Vegetation: Primary vegetation types, significant tree mapping, threatened or endangered species, habitat
- Soils: Prime farmland, healthy soils
- Human use: Views, proximity to transportation, proximity to diverse uses, adjacent properties
- Human health effects: The degree of air pollution on-site, the level of contamination present on-site, proximity to vulnerable populations

## SITE DESIGN

With the completion of the site assessment, the project team will have the necessary information to proceed with the site design. When designing the site, the entire development footprint should be considered as a whole, and the following strategies should be applied:

- Use native and/or adapted plants
- Remove invasive plants that exist on site
- Protect 100% of special-status vegetation on-site, as defined by local, state, or federal entities
- Eliminate up-lighting, glare, over-lighting, and light trespass (light spilling out of the project boundary) to conserve energy and avoid creating discomfort on adjacent properties
- Use reflective paving materials, which help distribute light across the site
- Protect on-site greenfield areas during construction and restore damaged habitat

- Provide exterior open space that promotes interaction with the environment, passive recreation, social interaction, and physical activity (according to LEED requirements, open spaces should contain vegetation and be accessible to all building users)
- Incorporate biophilic design to connect occupants with nature through natural elements, materials, and patterns
- Install bird-friendly glass on the building exterior to help reduce bird collisions

# MINIMIZING CONSTRUCTION IMPACTS

To minimize construction impacts, the project team should create and implement an erosion and sedimentation control (ESC) plan that follows the requirements of the U.S. EPA 2022 Construction General Permit (CGP), the EU Taxonomy: DNSH, Pollution Prevention, Item 4 (Noise and Dust), or a local equivalent. Key considerations include:

- The slope of the project site and where water will drain
- The total area and duration of ground disturbance, to identify air quality and rainwater runoff effects on neighboring properties
- The location of existing rainwater management systems that must be protected
- Weather and soil conditions that can create rainwater runoff or generate dust
- Construction entrances and their effects on local roads servicing the project site

To limit erosion and sedimentation, the following methods can be considered:

- Mulching
- Sediment (silt) fencing
- Erosion control blankets
- Berms and constructed ponds
- Seeding
- Straw bales

# RAINWATER MANAGEMENT

Green buildings should manage rainwater on-site to minimize impervious surfaces, slow runoff, and prevent overwhelming the municipal storm sewer system:

- Minimize impervious surfaces and use strategies such as vegetated roofs, rain gardens, porous pavements, grid pavers, and additional landscaping to allow rainwater to absorb into the soil
- Use pervious or open-grid paving in parking lots instead of asphalt or concrete
- Implement rainwater harvesting by collecting and filtering rainwater for reuse in irrigation, toilet and urinal flushing, or cooling tower process water, with cisterns used for storage

Another efficient rainwater management strategy is the use of **low-impact development (LID)**, which is an approach to mimic natural systems and to manage the stormwater closest to its source. LID strategies include decreasing impervious surfaces and increasing vegetation on-site.

# SUMMARY OF STRATEGIES TO ADDRESS WATER EFFICIENCY

## INDOOR WATER USE

When approaching the strategies to reduce the water consumption of a project, it is necessary to first calculate a **baseline** water usage and then find innovative ways to establish water use reductions from the baseline during the design phase. To calculate the baseline water usage of a project, first the baseline rates for flow and flush fixtures and fittings should be determined. The occupant usage should then be calculated to find out how much those fixtures and fittings will be used. Finally, project teams will need to discover innovative strategies for reducing water consumption as much as possible.

Indoor potable water consumption can be reduced by using water-efficient fixtures and fittings and/or using nonpotable water sources where appropriate. The use of **WaterSense-**labeled products can be a good strategy that will contribute to reducing indoor water usage. WaterSense is a program developed by the U.S. Environmental Protection Agency to identify high-performance, water-efficient fixtures and fittings.

Efficient fixtures and fittings for reducing potable water consumption include:

- Dual-flush toilets: This type of toilets have two buttons, one for full flush, the other one for half flush, depending on the type of waste.
- High-efficiency toilets (HETs): WaterSense-labeled toilets, which save 20% of the water typically used in every flush.
- Waterless urinals: These types of urinals do not need any water and have lower maintenance costs.
- Composting toilet systems: Such systems do not require any water since they treat waste with a microbiological process.
- Low-flow showerheads and faucets
- Faucets with low-flow aerators and/or motion sensors

To address indoor water use, LEED requires all projects to meet minimum efficiency requirements for fixtures and fittings by complying with the maximum specified flush and flow rates or by reducing the project's aggregate water use by at least 20% through the use of more efficient fixtures and fittings.

For residential and commercial clothes washers, residential dishwashers, and ice machines, LEED requires project teams to use ENERGY STAR–labeled equipment or equivalent performance.

In addition to high-efficiency fixtures, the use of harvested rainwater and/or graywater as the flush water of toilets can decrease potable water consumption.

**Graywater** is the untreated household water that does not come into contact with toilet waste. Used water from bathtubs, showers, bathroom washbasins, and clothes washers and laundry tubs can be examples of graywater and may be used as flush water in toilets or urinals.

Reducing **process water** usage is also essential in reducing potable water usage which is the type of water used by mechanical or other types of systems in buildings such as cooling towers or medical equipment in hospitals. Using reclaimed water, harvested rainwater, or graywater as process water can make a great contribution to reducing potable water consumption. **Reclaimed water** is former blackwater that has been treated and purified for reuse and can even be used for irrigation. (**Blackwater** is the term to describe the used water that has come into contact with waste. Thus, the water collected from the urinals and toilets can be classified as blackwater.)

If the projects also choose to install **closed-loop** systems, which use and circulate the same process water inside, instead of using **open-loop** systems, then even more potable water can be saved. Open-loop systems require outside water (which is called the **makeup water**) to support the system.

When doing water-use calculations for LEED, the water usage of **flush fixtures** such as toilets and urinals is measured in **gallons per flush** (**gpf**). For **flow fixtures**, such as sink faucets, showerheads, and aerators, the water usage is measured in **gallons per minute** (**gpm**).

# OUTDOOR WATER USE

The projects that contain landscaping will need to calculate the baseline **landscape water requirement (LWR)**. LWR is the amount of water that the landscape will require during the site's peak watering month. Next, project teams will need to discover innovative ways for reducing this value.

Water reductions in irrigation can be established through water-wise landscaping designs, water-efficient irrigation technologies, nonpotable water use (such as using reclaimed water, graywater, or harvested rainwater), and installation of submeters to track and log irrigation trends.

Following are the strategies to reduce outdoor water use:

- Xeriscaping, use of native and adapted plants, removal of invasive plants, and mulching to keep roots cool and prevent evaporation
- Limiting turf and grass areas, which require high irrigation
- Using high-performance irrigation systems such as drip irrigation systems (the most water-efficient, with short payback periods), bubbler distribution systems, and channeling water directly to root systems
- Using weather-based irrigation controllers to avoid watering during rain
- Scheduling irrigation in the early morning, the coolest part of the day
- Using alternative water sources such as reclaimed water, graywater, and harvested rainwater

For landscapes without permanent irrigation needs, LEED permits temporary irrigation during a two-year establishment period.

All projects with irrigation needs must reduce baseline outdoor water use by 30% through plant selection and irrigation system efficiency.

## WATER METETING

Submeters help projects track water use trends, evaluate fixture performance, find ways to reduce potable water consumption, identify high-consumption systems, and detect leaks early.

USGBC requires all projects to:

- Install or use existing permanent water meters to monitor, record, and report total water consumption for potable and alternative water sources (building and grounds).
- Share whole-project water use data with USGBC at least annually for five years, or until ownership/lessee changes.

Submetering individual building systems lets facility managers pinpoint major water uses and target reductions, while leak detection systems help managers and tenants catch leaks early to prevent waste and damage.

# SUMMARY OF STRATEGIES TO ADDRESS ENERGY AND ATMOSPHERE

This credit category contains the **highest total points** among the LEED credit categories (**33 points**) because energy use and the burning of fossil fuels are the biggest contributors to global warming.

## ENERGY DEMAND

Following are some strategies that can result in reducing the energy demand of buildings:

- **Building orientation** has the biggest effect on energy demand. By maximizing the southern exposure to the sun, buildings can have increased daylight and passive solar heating. (An example may be to orient the building in order to locate most of the windows facing south.) By minimizing western exposure, annual cooling costs can be dramatically reduced.
- **Community planning** can support building configurations in order to minimize solar gain in summer and maximize it in the winter.
- **Adjacent buildings** can be helpful in providing shade or insulation, or in blocking winds.
- **Passive strategies** like daylighting, thermal mass, and natural ventilation can reduce the energy demand for lighting, heating, and cooling.
- **Efficient building insulation** can reduce energy use.
- **Educating occupants** on their patterns of energy consumption can reduce energy demand.
- **3D computer modeling**, particularly when combined with building information modeling (BIM) software on energy design, can help a lot in evaluating different design alternatives by showing their outcomes. In addition, they can show the best locations for placing windows, doors, and other items.
- In existing buildings, conducting an **energy audit** can highlight the building systems that are not operating efficiently.
- **Rightsizing** is another factor that will affect energy demand. The bigger the building, the higher the energy demand; therefore, projects should not be developed larger than their needs.

# USE OF REFRIGERANTS

**Refrigerants** are substances commonly used in air-conditioning and refrigeration systems. Refrigerants in building systems can deplete stratospheric ozone, a gas that protects human health and the environment by absorbing harmful ultraviolet (UV) radiation. The destructive potential of these chemicals is measured by both their ozone depletion potential (ODP) and global warming potential (GWP). Because hydrochlorofluorocarbon (HCFC) refrigerants are known to have ozone-depleting potential, LEED prohibits their use.

Refrigerants are most harmful when they leak from cooling and heating systems. If there is any problem with the seals, leaks are likely to occur. For that reason, LEED also requires all projects to check both new and existing refrigerant-containing equipment for leaks.

**Natural refrigerants** are generally much more environmentally friendly than conventional alternatives. These include carbon dioxide ($CO_2$), water ($H_2O$), ammonia ($NH_3$), air, and hydrocarbons such as propane, ethane, and butane. Their ozone depletion potential (ODP) is zero, and their global warming potential (GWP) is zero or very low.

Equipment that contains less than 0.5 pounds (225 grams) of refrigerant is exempt from LEED requirements.

# ENERGY EFFICIENCY

After reducing energy demand, the project team should look for ways to improve energy efficiency, which will result in doing the same amount of work with less energy.

- Use synergistic strategies through the integrated process, such as installing a high-insulation building envelope that allows for a smaller HVAC system.
- Address building envelope air leaks.
- Maximize R-value (a material's resistance to heat flow — the higher, the better) while ensuring proper installation to avoid weak points and limit thermal bridging (heat transfer through conductive materials in the envelope).
- Increase thermal mass using high-density materials like concrete or thicker exterior brick, which absorb and store heat slowly to stabilize indoor temperatures and reduce HVAC loads.
- Use natural ventilation in appropriate climates by providing adequate windows and openings to reduce energy use and supply fresh air.
- Select HVAC equipment with high operating efficiency.
- Use energy-recovery ventilation, where exhaust air conditions incoming fresh air — transferring heat to incoming air in cold weather and reversing the process in hot weather.
- Regularly check and clean air filters in mechanically ventilated buildings to prevent HVAC equipment from working harder to move air through ducts.
- Prioritize daylighting and reduce lighting energy use.
- Use LEDs for artificial lighting.
- Install lighting controls such as photosensors, timers, occupancy sensors, and advanced lighting controls to shut off lights when not needed.

- LEED requires every project to meet a minimum energy performance baseline by complying with ANSI/ASHRAE/IES Standard 90.1.

# ELECTRIFICATION

Electrifying building systems to avoid on-site combustion of fossil fuels can significantly contribute to decarbonization, since burning fossil fuels generates greenhouse gas (carbon) emissions. **Electrification** refers to replacing non-electric energy sources with electricity.

- Install **heat pumps**, an energy-efficient system that draws thermal energy from water, air, or the ground into a refrigerant and transfers it elsewhere, replacing fossil-fuel-based space heating and water heating with electric operation.
- Use electric induction cooking, which uses a magnetic field to heat pots and pans directly, eliminating on-site combustion emissions.
- Design **grid-interactive buildings** that are highly efficient and can adjust how and when they use electricity based on grid conditions, helping reduce pressure during peak demand, supporting renewable energy use, and contributing to decarbonization and resilience — since increased electrification places new demands and timing patterns on power grids.
- Participate in **demand response**, where utility companies send alerts (demand response or curtailment events) to customers who agree to adjust usage during peak periods — for example, delaying EV charging or rescheduling equipment to operate later in the day.
- Use on-site energy storage, where batteries store electricity from on-site renewables or from the grid during low-demand hours for later use when demand rises.
- Use **thermal energy storage**, which stores heating or cooling capacity (rather than electricity) in an insulated tank for later use.
- Pair electrification with renewable energy sources, since electrification alone does not guarantee decarbonization; the electricity used by buildings must increasingly come from renewables for the built environment to become truly low-carbon.

# RENEWABLE ENERGY

**Renewable energy** includes solar, wind, wave, biomass, and geothermal power, plus certain forms of hydropower. Renewable energy sources can be either on-site or off-site.

**On-site renewable energy production** involves systems that generate renewable energy, such as solar photovoltaic (PV) panels, which convert sunlight into electricity. **Solar thermal collectors** can also capture solar energy to heat air or water for uses such as space heating, water heating, and swimming pool heating.

Off-site renewable energy production can involve the following sources:

- **Wind energy:** Wind turbines convert the kinetic energy of moving air into electricity.
- **Hydropower:** Hydropower converts the energy of flowing or falling water into electricity.
- **Bioenergy:** Electricity can be generated from biomass- or biofuel-based resources such as wood waste, landfill gas, agricultural crops or residues, animal waste, and other organic materials.
- **Wave power:** Wave energy converters can convert the energy of ocean waves into electricity.
- **Tidal power:** Tidal power is a form of hydropower that converts the energy of tides into electricity.

**Off-site renewable energy** is procured through contracts with utilities or the purchase of renewable energy certificates (RECs). If the local utility provider sells renewable electricity, the project can purchase it directly. If not, project teams can buy **renewable energy certificates (RECs)**, which are tradable, intangible instruments representing the environmental and other non-power attributes of renewable electricity generation. One REC generally represents 1 megawatt-hour (MWh) of electricity generated from an eligible renewable resource and delivered to the grid.

# POWER RESILIENCE

Project teams should also consider what will happen if the grid fails during an emergency. Power outages often occur during severe weather events or other disruptions, when access to electricity may be especially important for safety and communication. For this reason, projects should plan for a reliable backup power strategy.

**Renewable backup power** can offer significant advantages over conventional fossil-fuel generators. Systems such as solar panels or wind turbines are not dependent on fuel deliveries and can continue producing energy during longer outages if properly designed.

# BUILDING COMMISSIONING AND PERFORMANCE VERIFICATION

LEED requires all projects to implement building **commissioning**, a systematic process carried out by skilled personnel to verify that a building's systems are designed, installed, and operating in accordance with the owner's project requirements (OPR), project goals, and design specifications. Commissioning ensures that the project is designed and built as truly intended. Commissioning should begin early in design by specifying the OPR, which is then incorporated into the building design and implemented during construction.

By the end of the design development phase, a qualified commissioning provider (CxP) must be designated:

- The CxP must have direct commissioning experience on at least two projects of equal or larger scope and complexity.
- The CxP must be independent of the design and construction team and report directly to the owner.
- The CxP can be a third-party consultant, an owner's employee meeting the qualifications, or an employee of the design/construction firm if completely independent from the project's design or installation.
- If the CxP's employer has direct influence on design and construction, the commissioning plan must address potential conflicts of interest and demonstrate the CxP acts solely on behalf of the owner.

The CxP's responsibilities include:

- Assisting in developing the OPR
- Reviewing the BOD for compliance with the OPR
- Developing and implementing a commissioning (Cx) plan
- Confirming Cx requirements are incorporated into construction documents
- Developing functional performance tests
- Verifying system performance
- Preparing a final Cx report and documenting findings to the owner

The **Cx Plan** — a living document updated throughout the project that forms the basis of the final Cx report — will contain goals and objectives, general project information, systems to be commissioned, team member roles and responsibilities, key tasks and milestones, OPR and BOD reviews, functional performance testing procedures, verification of system performance, and reporting deficiencies and resolving issues.

During construction, the CxP will:

- Review submittals and substitutions for design deviations that impact the OPR
- Attend milestone meetings at 50% and 100% completion
- Perform a sample review of at least 10% of completed contractor documentation for quality assurance
- Witness functional performance testing of building elements and systems to verify installation matches specifications and design
- Include building envelope testing in the commissioning documents

The final Cx report, compiled by the CxP, will contain a Cx process overview, OPR and BOD, design review logs, approved submittals, Cx specifications, functional performance tests, and an issues and resolution log identifying closed items and proposed resolutions for outstanding items.

Finally, the CxP should develop an ongoing commissioning plan to keep systems operationally efficient throughout the building's life, providing facility managers with procedures, blank functional performance tests, and a recommended schedule for ongoing Cx activities.

Building systems should be monitored during operations to catch construction defects or equipment malfunctions that may undermine performance goals. Two strategies support this: monitoring-based commissioning (MBCx) and retrocommissioning.

**Monitoring-based commissioning** (**MBCx**) is a continuous, software-based process that monitors real-time data from building automation systems and meters. It enables operators to detect faulty equipment, identify unusual energy consumption patterns, and address issues as they occur. MBCx is led by a **monitoring-based commissioning provider** (**MBCxP**), either a contracted third party or a qualified in-house staff person.

**Retrocommissioning**, on the other hand, is a point-in-time systematic process applied to existing buildings to analyze and optimize system performance. It includes testing and documenting measures needed to bring buildings back on track to meet or exceed original operational goals.

# ENERGY METERING AND REPORTING—PREREQUISITE

LEED references ANSI/ASHRAE/IES Standard 90.1 for energy metering and requires projects to:

- Track both building energy use by source and on-site renewable energy generation.
- Report monthly energy data to USGBC at least once a year, including total energy use by source, on-site renewable energy generation, and peak electrical demand.

# SUMMARY OF STRATEGIES TO ADDRESS MATERIALS AND RESOURCES

## CONSERVATION OF MATERIALS

Project teams should approach materials with a mindset of sufficiency, material efficiency, and material circularity. **Sufficiency** refers to measures and practices that avoid unnecessary demand for energy, materials, land, and water while still delivering human well-being within planetary boundaries. **Material efficiency** refers to using less material to perform the same function or achieve the same results. **Material circularity**, on the other hand, refers to a system in which resources are continuously reused and recycled, reducing dependence on virgin materials and minimizing waste.

**Rightsizing** the project is the first step toward conservation of materials. An example of rightsizing includes designing buildings that are not larger than really needed and will therefore require fewer materials and resources.

Reusing existing materials, salvaged materials, and especially existing buildings results in tremendous material savings. If the project is a major renovation of an existing building, the project team should look for ways to reuse existing materials such as framing, envelopes, walls, flooring, ceilings, and roofing.

Using salvaged materials will contribute to reduction in the demand for virgin materials. Some types of salvaged materials that can be used in new buildings include bricks, doors, windows, flooring, cabinets, furniture, and tiles.

**Adaptive reuse** strategies, which can also be called "designing for flexibility," should be well considered in the buildings that may need frequent changes in layouts or floor plans. For example, a flexible floor plan for a hospital building can allow simple changes to be made regarding the sizes of rooms due to the changes in medical equipment that occur alongside the rapid evolution of the technology.

Material savings can also come from simplifying interior finishes. In some projects, a polished concrete slab can serve as the finished floor without adding carpet or other flooring materials. This approach of intentionally reducing the total amount of materials used in a building without compromising its function or performance is known as **dematerialization**.

Another common source of construction waste is **overage**, which refers to the extra material ordered as a precaution that often goes unused. Minimizing overage through careful design and procurement is ideal, but when surplus materials remain, project teams should look into returning them to the manufacturer or donating them to a reuse outlet rather than sending them to a landfill.

According to the EPA, the best way to eliminate waste is through **source reduction**. Source reduction refers to the exact sizing of the materials to be produced through prefabrication, modular construction, or similar methods, so that no waste is generated on-site.

# ENVIRONMENTALLY PREFERABLE MATERIALS

When selecting materials to be used in the project, the project team should consider a material's entire life cycle or life cycles since the aim of a sustainable material should be to have infinite life cycles. After the useful life of a material ends in one building, the material should be used as another product or a part of another product in order to continue being used. In other words, the first goal here is to implement all the strategies to stop the material from going to landfills as waste.

The following are some features of environmentally preferable materials:

- Locally harvested/extracted and manufactured
- Sustainably grown and harvested
- Made from **rapidly renewable materials** (which can replenish within 10 years)
- Do not contain any toxins
- Manufactured in factories that support human health and workers' rights
- Long-lasting and reusable
- Contain recycled content
- Have intended end-of-life scenarios that avoid a landfill
- Low in embodied energy

# LIFE-CYCLE ASSESSMENT

To examine the environmental cost of a building material, we need to consider the carbon emissions resulting from the material's manufacturing, transportation, installation, and disposal—in other words, its embodied carbon.

**Life-cycle assessment** (**LCA**), which quantitatively examines all the environmental effects of a product—or even a building—over its entire life cycle, allows project teams to understand the background of building products. Project teams can evaluate the trade-offs between different materials and select those that are the best fit for both the project and the environment. In LCA, a cradle-to-grave approach, and even more comprehensively, a cradle-to-cradle approach, may be used, allowing both the total energy use and other environmental consequences resulting from the creation of a material to be calculated.

A **cradle-to-cradle assessment** analyzes a product's life cycle from the initial extraction or harvesting of its resources to its new life as another product or as part of another product. Thus, cradle-to-cradle products are waste-free products that can be recycled—because of their high recyclable content—and reused.

By contrast, a **cradle-to-grave assessment** analyzes a product's life cycle from the initial extraction or harvesting of its resources to the end of its life, resulting in waste. Since cradle-to-grave products cannot be recycled, they have only one life cycle and ultimately end up in landfills.

For exam purposes, it is also important to be aware of **cradle-to-gate assessment**, which some manufacturers choose to conduct in order to evaluate a product's partial life cycle—from resource extraction or harvesting to its manufacture as a product ready for sale at the factory gate.

LEED requires all projects to quantify and assess the embodied carbon of building materials to increase awareness of the impacts of material choices. To do so, it requires quantifying cradle-to-gate (A1–A3) embodied carbon emissions for each building material, which include emissions associated with raw material extraction (A1), transport to the manufacturer (A2), and manufacturing (A3).

Because building material decisions are most flexible early in a project, teams can make the most meaningful reductions in embodied carbon during schematic design and design development.

## MATERIAL INGREDIENTS AND THEIR EFFECTS ON HUMAN HEALTH

Building materials can contain substances that are harmful to human health. One of the biggest indoor air quality concerns is volatile organic compounds. **Volatile organic compounds** (**VOCs**) are chemicals that evaporate at room temperature and can be released from a wide range of building products, including paints, adhesives, sealants, flooring, and furniture. Composite wood products like plywood and particleboard are a particular concern because they are commonly manufactured with formaldehyde-based adhesives, making **formaldehyde** one of the most critical indoor pollutants to control.

Beyond VOCs, many building materials contain other hazardous substances that deserve attention. Heavy metals such as **lead** and **chromium** are found in plumbing, wiring, and roofing products. **PFAS**, often called "**forever chemicals**" because they do not break down in the environment, show up in certain paints and stain-resistant fabric coatings.

## PRODUCT TRANSPARENCY

Project teams should be careful about **greenwashing** when selecting building products. Greenwashing refers to presenting a product as more environmentally friendly than it actually is. To avoid this, environmentally preferable materials should provide accurate and transparent information and declarations to buyers.

One example is the **environmental product declaration** (**EPD**), which reports a product's environmental impacts based on life-cycle assessment data. An EPD describes the environmental effects of raw material extraction, energy use, chemical composition, waste generation, and emissions to water, air, and soil throughout the product's sourcing and manufacturing process. Similarly, a product with a **health product declaration** (**HPD**) provides disclosure about its material ingredients, potential chemical hazards, related concerns, and additional health information.

# WASTE MANAGEMENT AND REDUCTION

The construction industry generates huge amounts of waste, and green building projects should reduce landfill contributions through a waste management plan covering both construction and operations.

During construction, deconstruction salvages far more usable material than demolition, and the same applies during renovations. Whatever cannot be reused should be sorted on-site for recycling. LEED rewards projects that implement a construction and demolition materials management plan early, awarding points based on the percentage diverted from landfill and incineration. Diversion strategies include off-site salvage, source separation, mixed recycling, and manufacturer take-back programs. Common diverted materials include concrete, metals, wood, brick, cardboard, gypsum board, ceilings, furniture, and carpet. Hazardous waste, excavated soil, and land-clearing debris are excluded from calculations, and materials sent for alternative daily cover or incineration are treated as waste.

For building operations, waste reduction starts during design by considering which waste streams will be present, what local services exist, and how much effort each stream requires. After occupancy, facility managers should track waste by source and material type, then reduce it through internal reuse, composting, and recycling. Every LEED project must have a recycling program covering organics/food waste, mixed paper, corrugated cardboard, glass, plastics, and metals, and must safely collect, store, and dispose of batteries, mercury-containing lamps, and electronic waste. Core and Shell projects must also communicate the waste management system in the tenant guidelines.

# SUMMARY OF STRATEGIES TO ADDRESS INDOOR ENVIRONMENTAL QUALITY

## INDOOR AIR QUALITY

In order to increase the indoor air quality of a building, first, the sources of contaminants should be identified, and necessary measures should be taken to prevent them from reaching the indoors, which is also called **source control**.

In order to avoid poor indoor air quality, a project should do the following:

- **Set minimum air quality standards:** Comply with ASHRAE 62.1-2022 or local codes, and monitor indoor air quality since adequate outdoor air alone may not be sufficient in polluted areas.
- **Stop secondhand smoke:** Prohibit smoking indoors and within 25 feet of entrances; schools must ban it campus-wide. For residential projects, LEED requires partitioning of the rooms and also prohibits smoking in the common spaces. Additionally, it requires a blower door test to ensure that air cannot leak from smoking units to others.
- **Use non-emitting or low-emitting materials:** Limit VOCs (such as formaldehyde) in adhesives, sealants, paints, flooring, furniture, and similar products.
- **Protect the site during construction:** Develop and implement an indoor air quality (IAQ) plan following SMACNA guidelines to protect HVAC equipment from dust and moisture, contain pollutants. During construction, project teams should first avoid the use of permanently installed equipment when possible. Only if this is not possible should project teams maintain effective filtration, then replace all filtration media before occupancy and complete testing and balancing with clean filters installed.
- **Schedule construction activities to minimize occupant exposure:** Clear contamination before occupancy through an air quality test or a flush-out (supplying ample fresh air to remove contaminants).
- **Increase ventilation rates:** Provide enough fresh air without overventilating; use demand-controlled ventilation with $CO_2$ sensors. **Demand-controlled ventilation** automatically adjusts outdoor airflow based on real-time signals from sensors. When $CO_2$ levels rise, the system increases ventilation; when levels drop, it reduces ventilation. If the climate in the project location is appropriate, natural ventilation can be established by the opening of windows and doors, which would also reduce the building operating costs. In an appropriate

climate, projects can use mixed-mode ventilation, which employs both HVAC and natural ventilation.

- **Monitor carbon dioxide ($CO_2$):** Install monitoring equipment to set thresholds and trigger increased fresh air supply when carbon dioxide levels rise.
- **Implement green cleaning practices:** Develop a green cleaning program covering certified products, permitted chemicals, staff training, pest control, and efficient equipment; install entryway grates or mats to keep contaminants out.
- **Periodically maintain and replace air filters:** Without quality air filtration media, no building can establish good indoor air quality. The **minimum efficiency reporting value (MERV)** rates the air filters according to their success in removing particles from the air. Air filters with higher MERV values will remove more particles from the air, but more energy will be consumed in passing the air through them.
- **Reduce radon levels:** Test for radon, a radioactive gas linked to lung cancer, and take urgent action if detected.

# INCREASE OCCUPANT COMFORT

Building occupants need to feel comfortable and in control of their environment in order to be healthy, happy, and productive in their buildings. There are several factors that will affect the comfort of an occupant. These include the following:

- **Thermal comfort:** Goes beyond room temperature to include air movement and humidity; HVAC systems should provide even temperatures and avoid blowing directly on occupants.
- **Daylighting:** Increases occupant health, comfort, and productivity while reducing artificial lighting costs. Effective design considers building orientation, configuration, glass types, room layout, furnishings, and surface colors to maximize daylight without glare or excess heat. Spaces designed to use daylight as their primary lighting source are called daylit spaces.
- **Daylighting systems and design:** Use passive systems (windows, glass doors, skylights, light tubes, light shelves) or active systems (mechanical devices that track sunlight, though less effective on cloudy days). **Light tubes** transport sunlight from roof to interior through reflective pipes; **light shelves** are horizontal overhangs that reflect daylight into spaces. Clerestory windows, properly sized/located windows, perimeter cubicles with private offices in the core, and low partitions help distribute daylight. Use adjustable window shades to control glare, and install zoned daylight controls that dim or shut off electric lights when natural light is sufficient.
- **Views:** Maximize quality views through vision glazing (windows for exterior views), low or no cubicle partitions, building orientation and layout adjustments, locating private offices near the core, and removing view-blocking objects. Many view strategies also enhance daylighting.
- **Acoustics:** Control sound isolation between spaces and HVAC background noise to support health, comfort, communication, and productivity — especially important in schools, open offices, and hospitals.
- **Lighting controls:** Provide individual, adjustable lighting controls and occupancy sensors to improve comfort and reduce costs. Evaluate layout and intensity to avoid glare, and install task lighting for specific work areas.

- **Temperature controls:** Provide individual thermal comfort controls so occupants can set preferred temperatures; in open offices, options like radiating heat panels under desks can address individual differences.
- **Ergonomics:** Address office layouts, workstation design, furnishings, and equipment based on occupant needs, with worker training upon installation. Key principles include flexibility, versatility, fit, and postural change.
- **Inclusive design**: By considering the physical, sensory, and cognitive needs of building occupants, LEED encourages project teams to go beyond basic accessibility measures and create environments that support people with different physical abilities, ages, and needs. This is the idea behind **inclusive design** (also called **universal design**), which aims to make buildings accessible to people of all ages, abilities, and demographics.

# SUMMARY OF STRATEGIES TO ADDRESS PROJECT PRIORITIES

LEED projects span the entire globe. They exist in different climate zones, typologies, and population densities, and they face varying levels of water and energy availability—as well as different local challenges. On top of that, every project is unique: it serves different users, responds to different goals, and operates under different constraints. Therefore, not every project should chase the same sustainability checklist. This credit category was created to give projects more flexibility to pursue what matters most for their specific context and priorities, and to encourage teams to innovate and accelerate new sustainable building strategies.

Moreover, industry knowledge and technologies evolve quickly. In previous LEED versions, project teams had to wait for the next major rating system release before new ideas, metrics, and solutions could be formally recognized. With LEED v5, USGBC created the **Project Priorities Library**, where new credits can be added as they are developed.

Under this credit category, project teams can earn up to 9 points by selecting credits from the USGBC's Project Priorities Library. These credits are designed to recognize efforts that are especially meaningful based on location, project type, performance ambition, or innovation. There are five credit types under the USGBC's Project Priorities Library as follows:

1) Regional priority:

   The **Regional Priority credit** is designed to reward projects for focusing on what matters most in their specific location. To earn points under this credit, the project team should identify a geographically specific environmental, social equity, or public health priority for the project site and then demonstrate achievement of a credit from the main LEED credit categories that addresses that priority in a significant, measurable way. For example, if the project is in a water-critical location, the project team may target the Enhanced Water Efficiency credit and make substantial, measurable reductions in potable water use to directly address local water scarcity. In this case, in addition to receiving points from the credit, the project may earn an extra point by submitting it as a Regional Priority credit.

2) Project-type credits:

**Project-type credits** are designated by USGBC to address the specific, unique needs of a particular adaptation or building application.

For example, the Places of Respite credit is a project-type credit available only to healthcare projects. It aims to provide patients, staff, and visitors with the health benefits of the natural environment by creating outdoor places of respite on the healthcare campus.

3) Exemplary performance:

For eligible LEED credits, performance beyond the LEED credit requirements can earn projects **exemplary performance points**, which are generally achieved by doubling the credit requirements or reaching the next incremental percentage threshold.

For example, in the Electric Vehicles credit, projects can earn up to 2 points by installing electric vehicle supply equipment that meets the credit's criteria. If the project team doubles the highest electric vehicle supply equipment parking-point threshold, they will receive 1 additional point for achieving a level of exemplary performance. Don't forget that not all LEED credits are eligible for exemplary performance; moreover, eligible credits have different exemplary performance requirements.

4) Pilot credits:

Pilot credits are tested across all rating system types and credit categories and include credits proposed for the next version of LEED. Project teams can pursue any of these pilot credits under Project Priorities credit and earn points by providing USGBC with feedback on the credit. Based on the feedback USGBC receives from projects that pursue these credits, pilot credits may become official credits in the next LEED rating system.

5) Innovative strategy:

LEED encourages projects to go beyond established green building principles by implementing new and unique strategies that are not currently addressed in the LEED rating systems. If projects develop innovations that are not covered in the rating system and can demonstrate reduced environmental impacts, increased decarbonization, and improved social impacts, they may receive additional points under the Project Priorities credit. Alternatively, they may also choose to achieve innovation credits from the USGBC's Project Priority Library.

# PROJECT PRIORITIES—CREDIT

- 1–9 Points (New Construction)

## CREDIT INTENT

To incentivize the achievement of credits that respond to geographically sensitive or climate-adaptation priorities—spanning environmental impact, social equity, and public health. To encourage projects to innovate and accelerate new sustainable building strategies.

## CREDIT REQUIREMENTS

Project teams can earn a maximum of 9 points by achieving any combination of the following:

1) Regional Priority

Achieve the Regional Priority credit from the Project Priority Library.

**AND/OR**

2) Project-Type Credits

Achieve a project-type credit from the Project Priority Library.

**AND/OR**

3) Exemplary Performance

Achieve an exemplary performance credit from the Project Priority Library.

**AND/OR**

4) Pilot Credits

Achieve a pilot credit from the Project Priority Library.

**AND/OR**

5) Innovative Strategies

Achieve an innovation credit from the Project Priority Library, or demonstrate significant, measurable environmental performance using a strategy that is not addressed in the LEED v5 rating system. To use this pathway, the project team must identify all of the following:

- The intent of the proposed innovation credit
- Proposed requirements for compliance
- Proposed submittals to demonstrate compliance
- The design approach or strategies used to meet the requirements

# LEED AP—CREDIT

- 1 Point (New Construction)

## CREDIT INTENT

To promote the team integration required for LEED projects and, at the same time, streamline the application and certification process.

## CREDIT REQUIREMENTS

At least one principal participant on the design team must be a LEED AP with a BD+C specialty. A project cannot earn this credit if it employs a LEED AP O+M as a principal participant for a LEED BD+C project.

The selected LEED AP with specialty must hold an active credential at the time of the project's certification review.

Legacy LEED APs (LEED APs without a specialty) do not qualify for this credit.

# FINAL EXAM TIPS

## FINAL EXAM TIPS

- **Read carefully. Make sure you read all the questions and the choices very carefully. If a question seems to have more than one answer, make sure you thoroughly understand the question and pay special attention to the wording. It doesn't matter how well you know the exam content; if you don't read carefully, the exam can easily trick you into selecting the wrong answer.**
- **Passing score.** The LEED Green Associate exam is scored between 125 and 200. A score of **170 or higher is required to pass**. However, the exam contains both scored and unscored questions. Unscored questions are used by GBCI to collect performance data and potentially serve as future exam questions. Test-takers will not know which questions are scored and which are unscored. Because of this, there is no way to know exactly how many correct answers you need in order to pass.
- **Time management.** During the exam, you will have **2 hours** to answer **100 questions**, which means you have about **72 seconds on average per question**. If you are stuck between two very close answers, instead of spending significant extra time trying to decide, mark the question and come back to it later. Often you will find it much easier to answer such questions when you return to them with a fresh perspective.
- **The test itself may help you.** If you forget something and struggle to answer a question, note that questions or answer choices you see later in the exam can help you remember the right answer to a previous question.
- **Answer every single question.** Leaving a question blank will not award any points, whereas making your best guess may result in the correct answer.
- **Test early in the day.** Schedule your exam for the morning and get a good night's sleep the night before. The exam requires strong concentration, and taking it with a fresh mind is the easiest way to boost your score. Don't study on exam day — take the test with a clear and refreshed mind.
- **Be prepared for distractions.** If you are taking your exam at a testing center, be prepared for potential noise distractions. Even though most settings will be quiet, some may not be. These centers offer a variety of exams, so expect to see people entering and exiting the testing area throughout your exam.
- **Manage test anxiety.** If you feel anxiety during the exam, pause for a moment. Take a deep breath, hold it for five seconds, and then slowly release it. Repeat this process a few times if needed. Thinking about a very funny moment in your life also works well to de-stress before moving on. Without managing your anxiety, your mind will struggle to function effectively, making every question seem more difficult than it really is. If a challenging question heightens your anxiety, use this breathing exercise, flag the question for later review, and move on. You will often find that when you return to that question later, it will be much easier to answer.
- **Use all your time.** If you have extra time at the end of the exam, use it to review your answers. Start with the questions you flagged and those you feel most unsure about.

# APPENDIX A – SUMMARY OF ASHRAE STANDARDS

| Name of the Standard / Program | Keywords |
|---|---|
| **ASHRAE 52.2-2017** | Filtration requirements |
| **ASHRAE 55-2023** | Thermal comfort requirements |
| **ASHRAE Standard 62.1-2022** | Ventilation and outdoor air requirements |
| **ASHRAE 62.2-2022** | Residential ventilation and exhaust requirements |
| **ANSI/ASHRAE/IES Standard 90.1 / ASHRAE 90.1-2022** | Energy efficiency, commissioning, energy metering, thermal bridging, peak load reduction, energy simulation, and energy performance improvement |
| **ASHRAE Standard 202-2024** | Enhanced commissioning framework |
| **ASHRAE 241-2023** | Air cleaning and infection risk management |

# APPENDIX B – IMPORTANT STANDARDS AND PROGRAMS

| Name of the Standard / Program | Keywords |
|---|---|
| **Intergovernmental Panel on Climate Change (IPCC)** | Referenced for climate resilience assessment |
| **ANSI Consensus National Standard Guide 2.0** | Referenced for integrative process guidance |
| **U.S. Department of Agriculture** | Defines prime farmlands |
| **Federal Emergency Management Agency (FEMA)** | Designates flood hazard areas |
| **U.S. Endangered Species Act** | Defines endangered and threatened species |
| **NatureServe** | Defines and assigns conservation status ranks for species and ecosystems |
| **International Union for Conservation of Nature (IUCN) Red List** | Identifies endangered and threatened species |
| **Institute of Transportation Engineers Parking Generation Manual** | Provides base parking capacity ratios |
| **ENERGY STAR** | Sets EVSE and energy-efficient equipment requirements |
| **U.S. EPA 2022 Construction General Permit (CGP)** | Sets requirements for the erosion and sedimentation control (ESC) plan |

| Name of the Standard / Program | Keywords |
|---|---|
| American Bird Conservancy Threat Factor Database | Defines bird-friendly glass requirements |
| FORTIFIED Commercial High Wind and Hail Specific Design Requirements | Sets requirements for hail and wind resilient design |
| U.S. National Tsunami Hazard Mitigation Program | Sets requirements for tsunami-resilient design |
| National Wildfire Coordinating Group Standards for Mitigation | Guides wildfire protection measures |
| American Forests | Created the Tree Equity Score |
| Illuminating Engineering Society of North America (IESNA) | Develops lighting specifications |
| WaterSense | A program developed by EPA to identify high-performance, water-efficient fixtures and fittings |
| Green-e | The leading certification program for the green power generation in the United States |
| California Department of Public Health (CDPH) Standard | Referenced for VOC emissions evaluation criteria |
| Sheet Metal and Air Conditioning National Contractors Association (SMACNA) IAQ Guidelines | Describes the necessary control measures to be taken during construction to protect indoor air quality |

# CONCLUSION

First of all, I would like to congratulate you on finishing this book and personally thank you for choosing it for your LEED Green Associate exam preparation. I truly hope it has exceeded your expectations.

To truly evaluate and sustain our performance, we all need feedback. With that in mind, please share any comments about this book, your exam score, and/or any other opinions you would like to share with me. If you found this book helpful, spreading the word by leaving a review on your favorite book review site, such as Amazon, Goodreads, or others, would mean the world to me. Your review supports my work and helps guide other readers.

If you'd like to receive my latest articles, books, and courses, be sure to join my mailing list.

Thank you for being part of this journey. I wish you all the best in your exam and your career!

| Scan the following QR codes to: | |
|---|---|
| **Send your feedback to the author and/or subscribe to the author's mailing list** | **Follow the author on LinkedIn** |
| | |

# GLOSSARY

**40/60 Rule:** A method used to choose the appropriate rating system for the project if the project seems to fit under multiple rating systems.

**Active Daylighting:** A system that tracks and collects sunlight using mechanical devices.

**Active Transportation**: Human-powered mobility such as biking, walking, or wheeling.

**Adapted Plants:** Types of plants that do not occur naturally in a specific location; however, they can nonetheless adapt easily to the climate of the region.

**Adaptive Reuse (Designing for Flexibility)**: A design strategy that allows buildings to accommodate future changes in layouts or floor plans without major reconstruction, particularly useful in buildings that may need frequent modifications.

**Addenda**: The overarching term that encompasses all changes, improvements, and modifications made to a LEED rating system. Addenda includes four types of changes: corrections, clarifications, LEED interpretations, and alternative compliance paths (ACPs).

**Advisory Task Forces**: Short-duration groups designated by USGBC staff to informally advise on issues limited in scope during the LEED development process.

**Albedo:** A measure of surface reflectivity on a scale from 0 to 1, where 0 represents a black surface that absorbs all solar radiation and 1 represents a white surface that reflects all solar radiation.

**Alternative Compliance Paths (ACPs):** Provide additional options for meeting credit requirements for international projects that need to reference local or regional standards instead of United States–based standards. There are two types of ACPs: Regional ACPs, which are credit requirements that may optionally be applied in specific regions of the world, and pilot ACPs, which are paths USGBC is considering for inclusion in future versions of LEED and can optionally be used in place of existing LEED credits.

**Alternative Fuel Vehicles:** Vehicles that consume nongasoline, low-polluting fuels like hydrogen, electricity, propane, compressed natural gas, liquid natural gas, methanol, or ethanol.

**Alternative Fuel:** Low-polluting fuels like hydrogen, electricity, propane, compressed natural gas, liquid natural gas, methanol, or ethanol.

**Alternative Water Source:** Nonpotable water from on-site surfaces, or freshwater sources, such as graywater, on-site reclaimed water, collected rainwater, captured condensate, and rejected water from reverse osmosis systems. Water from public utilities is excluded.

**Arc**: The online platform where project teams register their projects for LEED certification, complete prerequisite and credit documentation, and submit applications to GBCI for certification review. It serves as the primary tool for managing the entire LEED certification process from start to finish.

**Basis of Design (BOD):** Describes the information necessary to accomplish the owner's project requirements, including system requirements, design criteria, standards, guidelines, etc., developed by the architect/engineer.

**Bio-Based Materials:** Products other than food that are biological products, renewable agricultural materials, or forestry materials derived from biomass. Plants and animals can be examples of bio-based materials; however, hide products such as leather and other animal skin materials are excluded in LEED calculations.

**Biofuel:** Fuels produced from organic material. Biofuel includes untreated wood waste, landfill gas, agricultural crops or waste, animal waste, and other types of organic waste.

**Bioswale:** A stormwater control feature that uses a combination of an engineered basin, soils, and vegetation.

**Blackwater:** Used water that has come into contact with waste. Thus, the water collected from the urinals and toilets can be classified as blackwater.

**Blowdown:** Removal of the cooling tower's water in order to minimize deposit of scales.

**Brownfield Site:** Previously developed sites that were contaminated with waste or pollution. A site that has an abandoned building with unknown contamination can also be classified as a brownfield site.

**BUG Rating Method:** A luminaire classification system that classifies a luminaire according to backlight, uplight, and glare.

**Building Automation System (BAS):** A computer-based monitoring system which can monitor, coordinate and control every individual building system.

**Building Exterior:** Defined as everything from the waterproofing membrane, inclusive of the waterproofing membrane.

**Building Footprint:** Describes the area that the building sits on.

**Building Interior:** Defined as everything inside the waterproofing membrane.

**Built Environment:** All the man-made surroundings that are needed for human activity, from roads, to buildings, to neighborhoods.

**Charrettes:** Intense workshops that are generally held at the beginning of the project and during project milestones. These workshops bring together the entire project team, facility manager, and all the stakeholders and outside experts/consultants. Charrettes allow brainstorming and collaboration among different disciplines and experiences.

**Clean Waste**: Materials that are left over from construction and demolition that are nonhazardous.

**Climate Risk**: The likelihood and potential severity of harm resulting from a climate-related hazard, taking into account the building's vulnerability, its occupants, and the surrounding context.

**Climate Risk Assessment**: A systematic evaluation of how identified climate hazards could affect a building's performance, safety, occupants, and the owner's investment over the project's expected service life.

**Closed System:** A system that does not produce any waste product by circulating the same medium continuously.

**Commingled Waste:** Single-streamed waste for recycling.

**Commissioning (Cx):** A systematic process carried out by skilled personnel to verify that a building's systems are designed, installed, and operating in accordance with the owner's project requirements (OPR), project goals, and design specifications. Building commissioning ensures that the project is designed and built as truly intended.

**Compact Development**: Development strategies that promote location-efficient neighborhoods with high development density, reducing greenhouse gas emissions and urban sprawl.

**Compressed Workweek**: A work arrangement in which employees work longer hours on fewer days in exchange for an additional day off during the workweek, reducing commuting frequency and associated fossil fuel usage.

**Consensus Committees**: Established by the USGBC Board of Directors to support the creation and updating of specific LEED rating systems. Each consensus committee is structured to represent different technical and market interests, and no single interest group may make up more than 25% of the voting membership.

**Conventional Irrigation:** Common system used for irrigation, such as irrigation through sprinkler heads above the ground.

**Cradle-to-Cradle:** The practice of designing materials to have "infinite" life cycles by recovering and recycling them into new products, creating a closed-loop system instead of waste.

**Cradle-to-Grave:** A life-cycle approach that evaluates a product from raw material extraction through manufacturing and use, all the way to final disposal.

**Cradle-to-Gate Assessment:** A partial life-cycle assessment that evaluates impacts from raw material extraction/harvesting through manufacturing, ending when the product leaves the factory gate (before transportation to the site, use, and end-of-life).

**Decarbonization**: The process of reducing or eliminating carbon emissions from building operations, embodied materials, refrigerants, and transportation. Decarbonization is the largest impact area in LEED v5, accounting for approximately 50% of available points.

**Demand-Controlled Ventilation**: A ventilation strategy that automatically adjusts outdoor airflow in real time using sensor signals. Most demand-controlled ventilation systems use indoor carbon dioxide levels as an indicator of occupancy—when carbon dioxide levels rise, the system increases ventilation, and when carbon dioxide levels drop, it reduces ventilation.

**Demand Response (DR):** An intentional reduction in the electricity usage in response to a demand response (DR) event or changes in the price of electricity.

**Demand Response Event (Curtailment Event):** The period that the utility company asks for a reduction in electricity usage from its program participants.

**Dematerialization**: The practice of intentionally reducing the total amount of materials used in a building without compromising its function or performance.

**Development Density**: A measure of the total building floor area or total number of dwelling units on a unit of land relative to the total buildable area on that land.

**Development Footprint:** Named for the sum of all the areas that are affected by the project's activity in the project site.

**District Energy System (DES):** A central energy conversion plant that provides thermal energy, shared by a group of buildings.

**Diverse Uses:** Publicly available businesses that provide daily need goods or services. According to USGBC, diverse uses do not include ATMs, vending machines, and touch screens.

**Diversion Rate:** Percentage of waste materials diverted from landfill.

**Drip Irrigation Systems:** Microirrigation systems that drip water to the roots of plants to minimize the use of irrigation water and fertilizers. They are the most water-efficient systems and have very short payback periods.

**Dry Ponds (Detention Ponds):** Excavated basins that temporarily hold and slow down stormwater during rain events, but remain dry most of the time. By detaining runoff, they reduce downstream flooding and allow water to infiltrate gradually into the ground as the pond drains.

**EDUCATION @USGBC:** Education portal of USGBC.

**Electric Induction Cooking**: A cooking method that uses a magnetic field to directly heat pots and pans, eliminating emissions from on-site combustion.

**Electrification**: The process of replacing non-electric energy sources with electricity.

**Embodied Carbon:** Measures the total greenhouse gas emissions associated with a product's entire life cycle—from raw material extraction and manufacturing to transportation, installation, and disposal.

**Embodied Energy:** The total energy consumed resulting from a product's manufacturing, transportation, installation, use, and disposal.

**Emergent Properties:** Emergence of certain properties in the systems as a result of interaction of individual elements.

**Emissivity (Infrared or Thermal Emittance):** A measure that shows how much heat or infrared radiation a material can shed back into the atmosphere.

**Energy-Recovery Ventilation**: A system in which exhaust air is used to condition incoming fresh air. During cold weather, heat from outgoing air is transferred to incoming air, while in hot weather, the process works in reverse to reduce the cooling load.

**Energy Use Intensity (EUI):** A measurement unit that describes the building's energy usage relative to its size.

**Environmental Justice**: The principle that every person deserves equal protection from environmental and health hazards, equal access to decision-making, and equitable access to a healthy and sustainable environment.

**Environmental Product Declaration (EPD):** Disclosure that reports a product's environmental impacts based on life-cycle assessment data. An EPD describes the environmental effects of raw material extraction, energy use, chemical composition, waste generation, and emissions to water, air, and soil throughout the product's sourcing and manufacturing process.

**Environmental Tobacco Smoke (ETS):** Smoke from another person's cigarette, cigar, pipe, or electronic smoking device. Breathing in ETS is known as passive smoking, secondhand smoking, or involuntary smoking.

**Erosion and Sedimentation Control (ESC) Plan:** A plan developed to prevent erosion, sedimentation, and stormwater pollution to the water bodies, wetlands, and the whole neighborhood.

**Evapotranspiration:** The return of water to the atmosphere through evaporation from soil surfaces and transpiration from plants.

**Exemplary Performance Points:** Additional points generally achieved by doubling the credit requirements or reaching the next incremental percentage threshold.

**Extended Producer Responsibility (EPR):** A product stewardship policy approach that holds companies producing consumer goods responsible for managing their own products and packaging when consumers are finished with them.

**Extensive Vegetated Roofs:** Types of roofs that do not include a variety of plants and require little maintenance. Their soil layer is thinner compared with the intensive roofs since they are more designed for the smaller-sized vegetation.

**Facility Manager:** The professional responsible for overseeing the day-to-day operations, maintenance, and management of a building to ensure it functions efficiently and meets the needs of its occupants.

**Fenceline Community:** A community located immediately adjacent to or in close proximity to a source of pollution, such as an industrial facility, highway, or waste site, and therefore disproportionately exposed to environmental and health hazards.

**Floor-to-Area Ratio (FAR):** A density measure calculated by dividing a building's total gross floor area by the total lot (site) area. For example, a 20,000 sq ft building on a 10,000 sq ft lot has an FAR of 2.0.

**Flush-Out:** The process of supplying large volumes of fresh air to the building before or during occupancy to remove contaminated air and establish the desired level of indoor air quality.

**Formaldehyde:** A hazardous volatile organic compound commonly found in adhesives used to manufacture composite wood products such as plywood and particleboard. It is one of the most critical indoor pollutants to control.

**Functional Entry:** Any building opening that is open and used by pedestrians during business hours.

**Gallons per Flush (gpf):** A water-use metric for flush fixtures (toilets and urinals) that shows how many gallons are used per flush.

**Gallons per Minute (gpm):** A water-use metric for flow fixtures (faucets and showerheads) that shows how many gallons flow per minute.

**Glare:** Excessive brightness, often from high-angle light, that can cause visual discomfort and reduce visibility.

**Global Forest Watch:** An online platform that international projects may use to understand tree coverage in the project neighborhood.

**Graywater:** Untreated household water that did not come into contact with toilet waste. Used water from bathtubs, showers, bathroom washbasins, and water from clothes washers and laundry tubs can be examples of graywater and may be used as a flush water in toilets or urinals. This definition can change depending on the local codes.

**Green Building:** The practice of creating structures and using processes that are environmentally responsible and resource-efficient throughout a building's life cycle, from siting to design, construction, operation, maintenance, renovation, and deconstruction. This practice expands and complements the classical building design concerns of economy, utility, durability, and comfort.

**Green Classroom Professional (GCP) Certificate:** Designed for educators and school facility professionals who want to gain a foundational understanding of green building principles in the context of a school setting, helping them identify what supports or impedes healthy, resource-efficient, and environmentally sustainable learning spaces.

**Green Cleaning:** The use of environmentally friendly products and cleaning practices to maintain buildings while minimizing harm to human health and the environment.

**Green-e:** The leading certification program for renewable energy generation in the United States.

**Green Infrastructure:** Infrastructure to direct the rainwater collected from the impervious surfaces to the vegetation and soil surfaces without routing them to the storm sewer system.

**Greenfield:** Land that has not been previously developed.

**Greenwashing:** Refers to the presentation of a product or a material as being more environmentally friendly than it actually is.

**Grid-Interactive Building:** A building that can adjust how and when it uses electricity depending on grid conditions, helping reduce pressure during periods of high demand, support renewable energy use, and contribute to decarbonization and resilience.

**Halons:** Chemicals used in fire suppression systems.

**Hard Cost:** Costs that directly pay for the physical construction of the building, including labor, construction materials, and equipment.

**Hazard Assessment:** An evaluation of which threats are most relevant to a project by analyzing climate projections alongside the specific conditions of the site.

**Health Product Declaration (HPD):** Disclosure that provides a product's material ingredients, list of potential chemicals, related concerns, and additional health information.

**Heat Island Effect:** The phenomenon in which dark-colored, nonreflective surfaces absorb heat during hot weather and release it into the atmosphere, raising local temperatures.

**Heat Pumps:** An energy-efficient system for heating and cooling spaces and for heating water, draws thermal energy out of water, air, or the ground into a refrigerant and then transfers that heat to another place.

**Impervious Surface:** A surface that contains less than 50% perviousness.

**Inclusive Design (Universal Design):** A design approach that aims to make buildings accessible to people of all ages, abilities, and demographics.

**Indoor Air Quality Management Plan:** A plan developed to protect the indoor air quality for construction workers and building occupants.

**Infill Sites (Infill Developments):** Sites in which at least 75% of the total area was either previously developed or already being used for other purposes in urban areas.

**Inherently Nonemitting Materials:** Materials with very low or no VOC content.

**Integrated Pest Management (IPM):** A sustainable approach that combines knowledge about pests, nature, pest prevention, and control methods that minimize pest infestation and damage while minimizing hazards to the building occupants, the property itself, and the environment.

**Integrated Process:** An approach that emphasizes early and ongoing coordination among the project team. Owners, architects, and engineers (structural, electrical, mechanical), along with facility managers and other key parties, work collaboratively starting in the predesign phase to align goals and make better decisions.

**Intensive Vegetated Roofs:** Green roofs with deeper growing media that can support a wider variety of plants (including larger plants), but typically require more structural support and maintenance than extensive roofs.

**Intergovernmental Panel on Climate Change (IPCC)**: The United Nations body responsible for assessing the science related to climate change.

**Invasive Plants:** Plants that spread and damage the environment by taking over the adjacent existing native and adapted plants.

**Islanding**: The capability of an on-site energy system to disconnect from the main grid and operate independently during an outage, allowing the building to safely use its own energy resources and continue providing essential services.

**Land Trust:** A nonprofit organization that works on conserving lands.

**Landscape Water Requirement (LWR):** The amount of water that the landscape of the site will require during the site's peak watering month.

**LEED AP with Specialty:** LEED credential, created for professionals with advanced knowledge in green building practices and specialized in a particular LEED rating system.

**LEED Campus Program:** A program designed for multiple buildings on a shared site under the control of a single entity. It allows project teams to document shared site-related credits once and apply them across all individual building certifications on the campus.

**LEED Combined Review:** A type of LEED certification review in which the documentation for all design and construction prerequisites/credits is submitted for review at the end of the construction phase (for LEED BD+C and LEED ID+C rating systems).

**LEED Fellow:** LEED credential created to designate the most exceptional professionals in the green building industry.

**LEED Green Associate:** LEED credential created for professionals with a proven, up-to-date understanding of green building principles and practices.

**LEED Green Rater:** Professional who provides in-field verification to LEED Residential projects.

**LEED Impact Categories:** The high-level outcomes that LEED projects are designed to support. Every LEED prerequisite and credit aligns with one or more impact categories: Decarbonization, Quality of Life, and Ecological Conservation and Restoration.

**LEED Interpretation**: A precedent-setting official answer to a formal inquiry about the technical details of a LEED prerequisite, credit, or MPR. Unlike other forms of guidance, LEED interpretations apply to all LEED projects, not just the one that requested it. LEED interpretations are administered by USGBC and published in the Addenda database.

**LEED Project Administrator**: The primary project contact for GBCI. The LEED Project Administrator is the team member who acts as a project manager, overseeing the LEED project and organizing team members for certain tasks, credits, or prerequisites.

**LEED Project Boundary:** Portion of the site submitted for LEED certification. Defined by the platted property line of the project, including all land and water within it.

**LEED Recertification:** Required for LEED O+M projects to maintain their certification every 3 years. Projects certified under other LEED rating systems do not require recertification.

**LEED Scorecard**: Allows project teams to select the credits they plan to pursue and determine whether the total expected points will be sufficient to achieve the targeted level of LEED certification.

**LEED Split Review:** A type of LEED certification review in which the design prerequisites/credits are submitted for review during the design phase, and both the additional design prerequisites/credits and all the construction prerequisites/credits are submitted at the end of the construction phase (for LEED BD+C and LEED ID+C rating systems).

**LEED Steering Committee (LSC)**: A standing committee established by the USGBC Board of Directors. The LSC sets the strategic direction for LEED development, ensures consistency among all LEED rating systems, and reviews recommendations from other committees to confirm alignment with the overall LEED framework.

**LEED Volume Program**: A certification approach for organizations planning to certify a large number of new construction buildings or interior spaces based on a common prototype, streamlining certification by focusing on similarities in building design and construction practices regardless of location.

**Leverage Point:** The point where any action taken in the system can bring about significant results.

**Life-Cycle Approach:** Evaluates the entire life cycle of a project, product, or service.

**Life-Cycle Assessment (LCA):** A method that quantitatively evaluates a product's environmental impacts across its life cycle. A cradle-to-grave, or cradle-to-cradle, approach is used in LCA, and the total energy use and other environmental consequences resulting from the creation of that material are calculated.

**Life-cycle Costing (LCC):** A method that estimates a product's total cost over its life, including initial cost, operation, maintenance, service/repairs, and end-of-life disposal.

**Light Shelves:** Horizontal, light-reflecting overhangs that are positioned to reflect the daylight into the desired area of the building.

**Light Tubes:** Reflective tubular devices that capture daylight at the roof and channel it into interior spaces to provide natural light deeper inside a building.

**Lighting Power Density (Connected Lighting Power)**: The amount of power used per square foot or square meter when all lighting is turned on. Appropriate levels vary by space type.

**Linear Approach:** A traditional project delivery approach where each team member works separately in sequence, then hands their work off to the next person, with limited collaboration across disciplines.

**Load Shedding:** The intentional action by a power utility to reduce the load on the power system in order to prevent a total failure of the system.

**Load Shifting:** Storing energy during off-peak hours and using it later during peak-demand hours to reduce peak loads and electricity costs.

**Location Efficiency**: A measure of how well a project's surroundings reduce the need for driving by providing easy access to transit, services, and amenities within walking or cycling distance.

**Low-Impact Development (LID):** A rainwater management strategy that mimics natural site hydrology by managing stormwater close to where it falls. LID strategies include decreasing impervious surfaces and increasing vegetation on-site. LID is used to slow down the flow of rainwater out of site and ensures that the rainwater does not get contaminated.

**Makeup Water:** Water used to replace the lost water in open systems.

**Masking Systems:** Equipment used to reduce the background noise in spaces.

**Material Circularity**: A system in which resources are continuously reused and recycled, reducing dependence on virgin materials and minimizing waste.

**Material Efficiency**: Using less material to perform the same function or achieve the same results.

**Minimum Efficiency Reporting Value (MERV):** A rating system that measures how effectively an air filter removes particles from the air. Air filters with higher MERV values will remove more particles from the air, but more energy will be consumed in passing the air through them.

**Minimum Program Requirements (MPRs):** The minimum standards/characteristics that every LEED project should meet to pursue LEED certification. MPRs state whether a project can pursue LEED certification or not. They provide guidance on the types of projects that are eligible for LEED certification, protect the integrity of the LEED program, and reduce the number of issues that arise during the certification process.

**Monitoring-Based Commissioning (MBCx):** The process of using software to monitor real-time data from building automation systems and meters, enabling building operators to detect faulty equipment, identify unusual energy consumption patterns, and address issues as they occur.

**Monitoring-Based Commissioning Provider (MBCxP)**: A contracted third party or qualified in-house staff person responsible for leading the monitoring-based commissioning process.

**Mulching:** A protective layer applied to the surface of soil that will help to keep the roots of the plants cool and therefore prevent evaporation.

**Native Plants (Indigenous Plants):** Plants that occur and develop naturally in a specific location.

**Natural Refrigerants:** Refrigerants that occur in nature's biological and chemical cycles without human involvement, such as carbon dioxide ($CO_2$), water ($H_2O$), ammonia ($NH_3$), air, and hydrocarbons such as propane, ethane, and butane.

**Negative Feedback Loops:** A feedback loop where a change causes a response in the opposite direction to bring the system back toward balance. If a room gets warmer than the set temperature, the thermostat will send a signal to the air conditioning, and the air conditioning will stop blowing warm air.

**Net-Zero Energy Project:** A project that produces at least as much renewable energy as it consumes on an annual basis.

**Nonpoint Source Pollution:** Type of pollution in which its source cannot be identified and which generally results from multiple sources.

**Nonpotable Water:** Water that does not meet the human consumption standards.

**Nonprocess Energy (Regulated Energy):** The energy consumed by the items that are used to condition spaces and maintain comfort and amenities for building occupants.

**On-Site Water Retention:** The practice of collecting and retaining rainwater that falls onto pervious surfaces on the project site.

**Open Systems:** Systems that constantly consume other items, use them, and produce waste at the end.

**Open-Grid Pavement:** A pavement system with at least 50% unbound.

**Operational Carbon:** Emissions produced by running the building on a daily basis, such as energy used for heating, cooling, lighting, and powering equipment.

**Overage:** Extra material ordered as a precaution during construction that often goes unused. Minimizing overage through careful design and procurement reduces waste.

**Owner's Project Requirements (OPR):** A document that outlines the owner's vision and performance requirements for a project, including the LEED objectives the project needs to accomplish.

**Particulate Matter (PM):** A mix of tiny solid particles and liquid droplets floating in the air, most of which can only be detected using a microscope. PM can carry contaminants such as heavy metals, mold spores, dust mites, combustion by-products, pest droppings, and chemicals from pesticides, cleaning products, or building materials.

**Passive Daylighting:** A system that collects sunlight using static, nonmoving elements such as windows, glass doors, skylights, light tubes, and light shelves.

**Passive Survivability:** A building's ability to keep occupants safe by maintaining critical life-support conditions even when services such as power, heating fuel, or water are lost.

**Pilot Credits:** Credits being tested across all rating system types and credit categories and include credits proposed for the next version of LEED.

**Pilot Alternative Compliance Paths (Pilot ACPs):** Compliance paths that USGBC is considering for inclusion in future versions of LEED, which can optionally be used in place of existing LEED credits.

**Plug Loads (Receptacle Load):** Represents the electrical use by all the equipment that is connected to the electrical system via electrical receptacles.

**Positive Feedback Loop:** A feedback loop where a change causes more change in the same direction—A produces B, and B produces more of A. For example, in an interest-earning savings account, the balance earns interest, and as the balance grows, it earns even more interest, which accelerates the growth further.

**Postconsumer Recycled Content:** The recycled content of a used material. For example, recyclable printer paper can be sent to recycling after being used and can become a part of new printer paper. Other types of materials with postconsumer recycled content can be aluminum cans, water bottles, most glass, wood and steel products, newspapers, and more.

**Potable Water:** Water that is approved for human use that meets or exceeds US Environmental Protection Agency drinking water quality standards (or a local equivalent outside the United States.)

**Preconsumer Recycled Content:** The content of a material that is recycled before reaching any consumer. An example is sawdust generated during the manufacturing of a wood product that is recycled to be used inside an MDF board (medium-density fiberboard).

**Preferred Parking:** Parking spaces that are closest to the main entrance of a building.

**Prerequisites:** The minimum requirements that all buildings under a certain rating system must meet in order to achieve LEED certification.

**Previously Developed Site:** A site that contains at least 75% previously developed land.

**Prime Farmland:** Land that is used for producing food, feed, forage, fiber, and oilseed crops or is available for these uses, as determined by the U.S. Department of Agriculture's Natural Resources Conservation Service.

**Prius Effect:** The idea that people change their behavior when they can see real-time feedback. In other words, people can respond to something only if they have immediate information about it.

**Process Water:** Water used by mechanical or other types of systems in buildings such as cooling towers or medical equipment in hospitals.

**Project Priorities Library**: A USGBC resource where new credits can be added as they are developed.

**Project-Type Credits**: Credits designated by USGBC to address the specific, unique needs of a particular adaptation or building application, such as the Places of Respite credit available only to healthcare projects.

**Radon:** A radioactive gas naturally found in soils, rocks, and water bodies. Exposure to radon is harmful to human health and is the second leading cause of lung cancer after smoking.

**Rainwater Harvesting:** An aspect of rainwater management that collects and filters the rainwater to be reused as an alternative to potable water.

**Rapidly Renewable Materials:** Natural materials that can replenish within 10 years.

**Reclaimed Water:** Former blackwater that has been treated and purified for reuse.

**Reference Soil:** The site's original, undisturbed native soil—the soil condition that existed before development and that is used as the baseline for comparison (for example, when evaluating soil restoration or soil health).

**Refrigerant:** Substances used to transfer heat.

**Regenerative Design:** A design approach where the project is intended to give back more than it takes—producing more energy than it consumes, generating little to no waste, and actively restoring ecosystems and habitats that development has degraded. Excess energy produced on site can be returned to the utility grid, making the building a contributor to the energy supply rather than just a consumer.

**Regional Alternative Compliance Paths (Regional ACPs)**: Credit requirements that may optionally be applied in specific regions of the world to address local standards and conditions.

**Renewable Energy Certificates (RECs):** Tradable, intangible instruments representing the environmental and other non-power attributes of renewable electricity generation. One REC generally represents 1 megawatt-hour (MWh) of electricity generated from an eligible renewable resource and delivered to the grid.

**Renewable Energy:** A type of energy that is derived from renewable sources. Renewable energy includes solar, wind, wave, biomass, and geothermal power, plus certain forms of hydropower.

**Resilience:** A building's ability to prepare for, withstand, respond to, and recover from disruptive events.

**Retrocommissioning:** A systematic process for analyzing and optimizing building system performance involving a commissioning provider. It includes testing of building systems and documentation of any measures needed to improve performance.

**Reverberation Time:** The time span between when a sound is produced and when it dies away.

**Scope 1 Energy:** Direct energy generated from sources owned or controlled by the organization, such as on-site fuel combustion.

**Scope 2 Energy:** Indirect energy purchased from external sources, such as electricity, steam, or district heating and cooling.

**Scope 3 Energy:** Indirect energy from sources not owned or directly controlled by the organization, such as employee commuting, business travel, or supply chain activities.

**Sensitive Lands:** Ecologically sensitive areas such as prime farmland, floodplain, habitat, water bodies, or wetland.

**Site Assessment:** A part of the integrative process, which clearly shows the project teams the properties of the site, including its topography, hydrology, climate, soil types, water availability, and human health effects.

**Smart Growth:** A neighborhood development approach that protects undeveloped lands and contributes to project development in locations near jobs, schools, shops, and other diverse uses.

**Social Determinants of Health:** The conditions in which people are born, live, and work that play a significant role in shaping well-being, including factors such as access to clean air, safe housing, green spaces, nutritious food, and quality education.

**Soft Cost:** Costs needed to plan, design, and manage a project that do not physically become part of the building. Soft costs include items such as design and consulting fees, project management and supervision, permits, legal fees, financing, insurance, and taxes.

**Solar Reflectance (SR):** Shows the solar energy that is reflected by a surface on a scale of 0 to 1. A black surface will have a SR of 0 while a white surface will have a SR of 1.

**Solar Reflectance Index (SRI):** Indicates a material's ability to stay cool by reflecting solar radiation and emitting thermal radiation. Thus, both the solar reflectance and emissivity of a material will be combined to rank the material.

**Source Control:** The practice of identifying sources of indoor contaminants and taking measures to prevent them from reaching the indoor environment.

**Source Reduction:** The exact sizing of materials through prefabrication, modular construction, or similar methods to prevent waste generation on-site. It also covers the use of products with less packaging.

**Suburban Sprawl:** The expansion of populations away from central urban areas into low-density areas.

**Sufficiency**: A set of measures and practices that avoid unnecessary demand for energy, materials, land, and water while still delivering human well-being within planetary boundaries.

**Sustainable Development:** Meeting the needs of the present without compromising the ability of future generations to meet their own needs.

**Systems Thinking:** Refers to the understanding of each and every system of a building while also understanding their relationships and looking at the project as a whole.

**Technical Advisory Groups (TAGs)**: Committees that advise the LEED Steering Committee and consensus committees on specific technical aspects of LEED, such as materials or indoor air quality.

**Thermal Bridging**: Areas where conductive materials within the building envelope permit unwanted heat transfer in both cold and hot weather conditions.

**Thermal Energy Storage**: A system that stores heating or cooling capacity for later use, typically in an insulated tank, rather than storing electricity.

**Thermal Mass:** The ability of a material to absorb and store energy. High-density materials such as concrete and brick require significant amounts of heat to change temperature.

**Tier 1 Renewable Energy**: On-site renewable energy generation, including systems located on the project site, on the surrounding campus, or on the site of a social impact project where the system is donated and ownership is transferred to the recipient. Eligible resources include solar electric (photovoltaic), solar thermal, wind, and recovered heat from municipal wastewater.

**Tier 2 Renewable Energy**: New off-site renewable electricity from generation assets contracted to be operational within two years of occupancy or within five years of their commercial operations date.

**Tier 3 Renewable Energy**: Off-site renewable electricity or fuels that are Green-e Energy certified or equivalent.

**Total Volatile Organic Compounds (TVOC):** The combined concentration of many VOCs reported as a single value.

**Transportation Demand Management (TDM)**: A set of rules and strategies designed to reduce single-occupancy vehicle trips, including measures such as public transit incentives, carpool programs, employer commuter benefits, limits on parking supply, and policies that encourage walking, biking, and public transit.

**Tree Equity Score**: A tool that highlights inequities in tree cover and identifies where trees are needed most in each community, helping address disparities by focusing resources in neighborhoods with the highest need.

**Triple Bottom Line (Three Pillars of Sustainability)**: A framework that evaluates a project's success across three dimensions: economic prosperity (profit), social equity and justice (people), and environmental stewardship (planet).

**Uplight**: Light emitted above the horizontal plane that contributes to sky glow.

**Volatile Organic Compounds (VOCs)**: Chemicals that evaporate at room temperature and can be released from building products such as paints, adhesives, sealants, flooring, and furniture. Some VOCs can cause short-term effects like headaches or long-term health problems including cancer.

**Walk Score**: A measure of a location's walkability on a scale of 0 to 100, based on walking routes to nearby amenities. A higher score indicates more amenities and services within a short walking distance.

**Wet Pond (Retention Pond):** Pond designed to hold a specific amount of water indefinitely.

**Working Groups**: Groups established by the LEED Steering Committee or a consensus committee to complete a specific task with a clearly defined scope and timeline.

**Xeriscaping:** Type of landscaping that does not need any irrigation.

**Zero-Lot Project:** A type of project in which the building footprint extends all the way to one or more property (lot) lines, leaving little or no space for setbacks on those sides, except for limited access areas (e.g., entryways, walkways, drive aisles, or service access).